Digital Video and DSP: Instant Access

The Newnes Instant Access Series

FPGAs: Instant Access
Clive Maxfield
ISBN-13: 978-0-7506-8974-8

Digital Video and DSP: Instant Access
Keith Jack
ISBN-13: 978-0-7506-8975-5

Digital Signal Processing: Instant Access
James D. Broesch
ISBN-13: 978-0-7506-8976-2

For more information on these and other Newnes titles visit: www.newnespress.com.

Digital Video and DSP: Instant Access

Keith Jack

AMSTERDAM • BOSTON • HEIDELBERG • LONDON
NEW YORK • OXFORD • PARIS • SAN DIEGO
SAN FRANCISCO • SINGAPORE • SYDNEY • TOKYO
Newnes is an imprint of Elsevier

Newnes is an imprint of Elsevier
30 Corporate Drive, Suite 400, Burlington, MA 01803, USA
Linacre House, Jordan Hill, Oxford OX2 8DP, UK

Recognizing the importance of preserving what has been written, Elsevier prints its books on acid-free paper whenever possible.

Library of Congress Cataloguing-in-Publication Data
Application submitted.

British Library Cataloguing-in-Publication Data
A catalogue record for this book is available from the British Library.

ISBN: 978-0-7506-8975-5

For information on all Newnes publications
visit our Web site at: www.books.elsevier.com

Printed in Canada
08 09 10 11 12 13 10 9 8 7 6 5 4 3 2 1

Typeset by Charon Tec Ltd., A Macmillan Company (www.macmillansolutions.com)

Contents

1. Video Overview

Video Definitions 1
Video Today 2
Video Data 3
Video Timing 4
Video Resolution 6
Video and Audio Compression 7
Other Video Applications 8
Standards Organizations 10

2. Color Spaces

Definitions 15
RGB Color Space 15
YUV Color Space 16
YIQ Color Space 17
YCbCr Color Space 18
HSI, HLS, and HSV Color Spaces 19
Chromaticity Diagram 22
Non-RGB Color Space Considerations 25
Gamma Correction 26

3. Video Signals

Definitions 31
Digital component video background 32
480i and 480p systems 35
576i and 576p systems 41
720p systems 48
1080i and 1080p systems 54

4. Video Interfaces

Analog video interfaces 63
Definitions 63
S-Video interface 64

SCART interface 65
SDTV RGB interface 65
HDTV RGB interface 67
SDTV YPbPr interface 69
HDTV YPbPr interface 71
D-Connector interface 75
Other Pro-Video analog interfaces 77
VGA interface 78
Digital video interfaces **78**
Digital video interface definitions 78
Pro-video component interfaces 81
Pro-video composite interfaces 94
Pro-video transport interfaces 105
IC component interfaces 107
Consumer component interfaces 110
Consumer transport interfaces 119

5. **Digital Video Processing**

Processing definitions **125**
Display enhancement **126**
Video mixing and graphics overlay **130**
Luma and chroma keying **131**
Video scaling **132**
Scan rate conversion **140**
Noninterlaced-to-interlaced conversion **143**
Interlaced-to-noninterlaced conversion **144**
DCT-based compression **147**

6. **NTSC, PAL and SECAM**

Definitions **151**
NTSC overview **151**
PAL overview **156**
SECAM overview **160**
Enhanced television programming **162**

7. **MPEG-1, MPEG-2, MPEG-4 and H.264**

MPEG Definitions **165**
MPEG-1 **166**
MPEG-2 **174**
MPEG-4 **187**
MPEG-4.10 (H.264) VIDEO **195**

8. Digital TV

Definitions 201
ATSC digital television 201
OpenCable™ digital television 210
DVB digital television 213
ISDB digital television 216

Index 223

Chapter 1

Video Overview

In an Instant

- Video Definitions
- Video Today
- Video Data
- Video Timing
- Video Resolution
- Video and Audio Compression
- Other Video Applications
- Standards Organizations

Video Definitions

Although there are many variations and implementation techniques, video signals are just a way of transferring visual information from one point to another. The information may be from a VCR, DVD player, a channel on the local broadcast, cable television, or satellite system, the Internet, cell phone, MP3 player, or one of many other sources. Invariably, the video information must be transferred from one device to another. It could be from a satellite set-top box or DVD player to a television. Or it could be from one chip to another inside the satellite set-top box or television. Although it seems simple, there are many different requirements, and therefore many different ways of doing it.

This book will cover the engineering essentials of this important technology. First we'll define some video terms and concepts.

A *color space* is a mathematical representation for a color. Initially video contained only gray scale, or black-and-white, information. When color broadcasts were being developed, attempts were made to transmit color video using RGB (red, green, blue) color space data, but that technique occupied too much bandwidth so other alternative color spaces were developed. They will be covered in more detail in Chapter 2.

Component video is video using three separate color components, such as YCbCr (digital), YPbPr (analog), or R′G′B′ (digital or analog). *Composite video* uses a single signal to contain color, brightness and timing information.

Compression is an important part of video technology. *MPEG* stands for Moving Picture Experts Group, an international standards group that develops various compression algorithms. MPEG video compression takes advantage of the redundancy on a frame by frame basis of a normal video sequence. There are several different MPEG standards which we'll cover in later chapters.

VIDEO TODAY

A few short years ago, the applications for video were somewhat confined—analog video technology was used for broadcast and cable television, VCRs, set-top boxes, televisions, and camcorders. Since then, there has been a tremendous and rapid conversion to digital video, mostly based on the MPEG-2 video compression standard.

The average consumer now uses digital video every day thanks to continuing falling costs. This trend has led to the development of DVD players and recorders, digital set-top boxes, digital television (DTV), portable video players, and the ability to use the Internet for transferring video data. Equipment for the consumer has also become more sophisticated, supporting a much wider variety of content and interconnectivity. Today we have:

- *HD DVD and Blu-ray Players and Recorders*. In addition to playing CDs and DVDs, these advanced HD players also support the playback of MPEG-4.10 (H.264), and SMPTE 421 M (VC-1) content. Some include an Ethernet connection to enable content from a PC or media server to be easily enjoyed on the television.
- *Digital Media Adapters*. These small, low-cost boxes use an Ethernet or 802.11 connection to enable content from a PC or media server to be easily enjoyed on any television. Playback of MPEG-2, MPEG-4.10 (H.264), SMPTE 421 M (VC-1), and JPEG content is typically supported.
- *Digital Set-Top Boxes*. Cable and satellite set-top boxes are now including digital video recorder (DVR) capabilities, allowing viewers to enjoy content at their convenience. Use of MPEG-4.10 (H.264) and SMPTE 421 M (VC-1) now enables more channels of content and reduces the chance of early product obsolescence.
- *Digital Televisions (DTV)*. In addition to the tuners and decoders being incorporated inside the television, some also include the digital media adapter capability. Support for viewing on-line video content is also growing.
- *IPTV Set-Top Boxes*. These low-cost set-top boxes are gaining popularity in regions that have high-speed DSL and FTTH (fiber to the home) available. Use of MPEG-4.10 (H.264) and SMPTE 421 M (VC-1) reduces the chance of early product obsolescence.
- *Portable Media Players*. Using an internal hard disc drive (HDD), these players connect to the PC via USB or 802.11 network for downloading a wide variety of content. Playback of MPEG-2, MPEG-4.10 (H.264), SMPTE 421 M (VC-1). and JPEG content is typically supported.
- *Mobile Video Receivers*. Being incorporated into cell phones, MPEG-4.10 (H.264) and SMPTE 421 M (VC-1) is used to transmit a high-quality video signal. Example applications are the DMB, DVB-H and DVB-SH standards.

There are many engineering challenges faced when incorporating video into today's product designs. Implementing real-world solutions is not easy, and many engineers have little knowledge or experience in this area. This book is a quick-start guide for those charged with the task of understanding and implementing video features into next-generation designs.

VIDEO DATA

Initially, video contained only gray-scale (also called black-and-white) information. While color broadcasts were being developed, attempts were made to transmit color video using analog RGB (red, green, blue) data. However, this technique occupied three times more bandwidth than the gray-scale solution, so alternate methods were developed that led to using Y, R–Y, and G–Y data to represent color information, where Y represents the *luma* (black-and-white part), and R–Y and G–Y represent color difference signals made by subtracting the Y from the red and blue components. A technique was then developed to transmit this Y, R–Y, and G–Y information using one signal, instead of three separate signals, and in the same bandwidth as the original gray-scale video signal.

Today, even though there are many ways of representing color video (R′G′B′, YIQ, YCbCr, YPbPr, YUV, and others are covered in Chapter 2), they are still all related mathematically to RGB.

S-Video was developed for connecting consumer equipment together (it is not used for broadcast purposes). It is a set of two analog signals, one gray-scale (Y) and one that carries the analog R–Y and B–Y color information in a specific format (also called C or chroma). Once available only for S-VHS, it is now supported on most consumer video products.

Although always used by the professional video market, analog RGB video data has made a temporary comeback for connecting high-end consumer equipment together. Like S-Video, it is not used for broadcast purposes.

Insider Info

A variation of the Y, R–Y, and G–Y video signals, called YPbPr, is now commonly used for connecting consumer video products together. Its primary advantage is the ability to transfer high-definition video between consumer products. Some manufacturers incorrectly label the YPbPr connectors YUV, YCbCr, or Y(B-Y) (R-Y).

Digital Video

The most common digital signals used are RGB and YCbCr. RGB is simply the digitized version of the analog RGB video signals. YCbCr is basically the digitized version of the analog YPbPr video signals, and is the format used by DVD and digital television.

Technology Trade-offs

There is always the question of "What is the best connection method for equipment?" For DVD players and digital cable/satellite/terrestrial set-top boxes, the typical order of decreasing video quality is:

1. HDMI (digital YCbCr)
2. HDMI (digital RGB)
3. Analog YPbPr
4. Analog RGB
5. Analog S-Video
6. Analog Composite

Some will disagree about the order. However, most consumer products do digital video processing in the YCbCr color space. Therefore, using YCbCr as the interconnect for equipment reduces the number of color space conversions required. Color space conversion of digital signals is still preferable to D/A (digital-to-analog) conversion followed by A/D (analog-to-digital) conversion, hence the positioning of HDMI RGB above analog YPbPr. The computer industry has standardized on analog and digital RGB for connecting to the computer monitor.

VIDEO TIMING

Although it looks like video is continuous motion, it is actually a series of still images, changing fast enough that it looks like continuous motion, as shown in Figure 1.1. This typically occurs 50 or 60 times per second for consumer video, and 70–90 times per second for computer displays. Special timing information, called *vertical sync*, is used to indicate when a new image is starting.

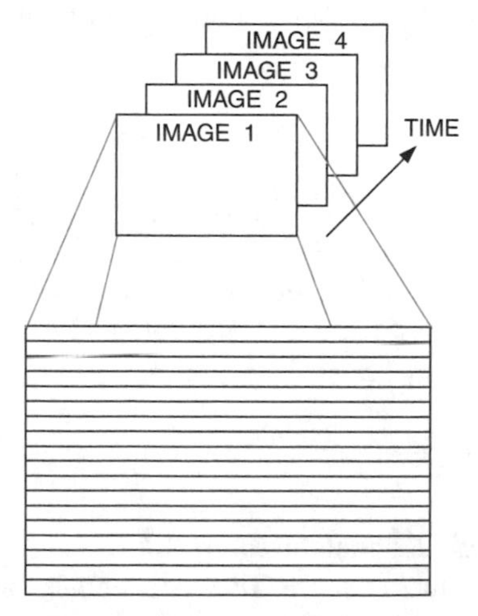

FIGURE 1.1 Video is composed of a series of still images. Each image is composed of individual lines of data.

Each still image is also composed of *scan lines*, lines of data that occur sequentially one after another down the display, as shown in Figure 1.1. Additional timing information, called *horizontal sync*, is used to indicate when a new scan line is starting.

The vertical and horizontal sync information is usually transferred in one of three ways:

1. Separate horizontal and vertical sync signals
2. Separate composite sync signal
3. Composite sync signal embedded within the video signal

The composite sync signal is a combination of both vertical and horizontal sync.

Insider Info

Computer and consumer equipment that uses analog RGB video usually uses technique 1 or 2. Consumer equipment that supports composite video or analog YPbPr video usually uses technique 3. For digital video, either technique 1 is commonly used or timing code words are embedded within the digital video stream.

Interlaced vs. Progressive

Since video is a series of still images, it makes sense to simply display each full image consecutively, one after another.

This is the basic technique of *progressive*, or noninterlaced, displays. For progressive displays that "paint" an image on the screen, such as a CRT, each image is displayed starting at the top left corner of the display, moving to the right edge of the display. The scanning then moves down one line, and repeats scanning left-to-right. This process is repeated until the entire screen is refreshed, as seen in Figure 1.2.

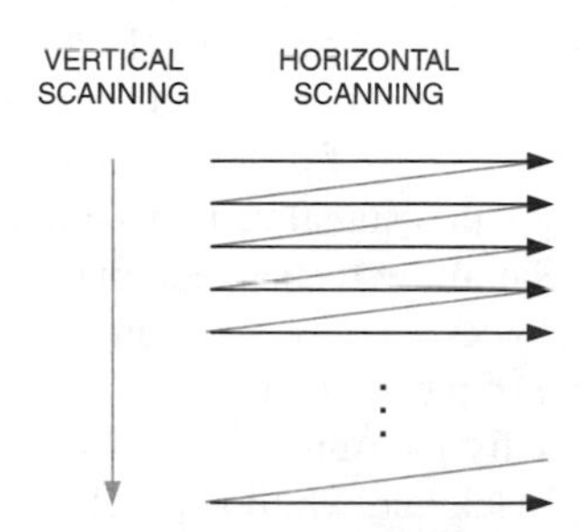

FIGURE 1.2 Progressive displays "paint" the lines of an image consecutively one after another.

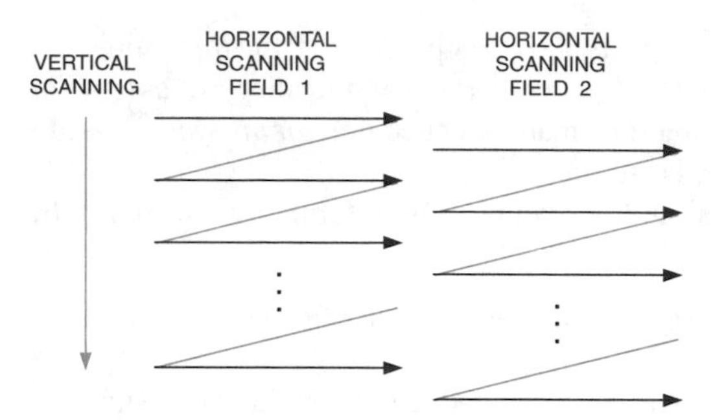

FIGURE 1.3 Interlaced displays "paint" first one-half of the image (odd lines), then the other half (even lines).

In the early days of television, a technique called *interlacing* was developed to reduce the amount of information sent for each image. By transferring the odd-numbered lines, followed by the even-numbered lines (as shown in Figure 1.3), the amount of information sent for each image was halved. Today, most broadcasts (including HDTV) are still transmitted as interlaced. Most CRT-based displays are still interlaced while LCD, plasma, and computer displays are progressive.

FAQs

Given the advantage of interlacing, why bother to use progressive?

With interlace, each scan line is refreshed half as often as it would be if it were a progressive display. Therefore, to avoid line flicker on sharp edges due to a too-low frame rate, the line-to-line changes are limited, essentially by vertically lowpass filtering the image. A progressive display has no limit on the line-to-line changes, so it is capable of providing a higher-resolution image (vertically) without flicker.

VIDEO RESOLUTION

Video resolution is one of those "fuzzy" things in life. It is common to see video resolutions quoted as "720 × 480" or "1920 × 1080." However, those are just the number of horizontal samples and vertical scan lines, and do not necessarily convey the amount of useful information.

For example, an analog video signal can be sampled at 13.5 MHz to generate 720 samples per line. Sampling the same signal at 27 MHz would generate 1440 samples per line. However, only the number of samples per line has changed, not the resolution of the content.

Therefore, video is usually measured using lines of resolution. In essence, this is how many distinct black and white vertical lines can be seen across the display. This number is then normalized to a 1:1 display aspect ratio (dividing the number by 3/4 for a 4:3 display, or by 9/16 for a 16:9 display). (Of course, this results in a lower value for widescreen (16:9) displays, which goes against intuition.)

Standard-Definition

Standard-definition video is usually defined as having 480 or 576 interlaced active scan lines, and is commonly called "480i" and "576i," respectively.

For a fixed-pixel (non-CRT) consumer display with a 4:3 aspect ratio, this translates into an active resolution of 720 × 480i or 720 × 576i. For a 16:9 aspect ratio, this translates into an active resolution of 960 × 480i or 960 × 576i.

Enhanced-Definition

Enhanced-definition video is usually defined as having 480 or 576 progressive active scan lines, and is commonly called "480p" and "576p," respectively.

For a fixed-pixel (non-CRT) consumer display with a 4:3 aspect ratio, this translates into an active resolution of 720 × 480p or 720 × 576p. For a 16:9 aspect ratio, this translates into an active resolution of 960 × 480p or 960 × 576p.

The difference between standard and enhanced definition is that standard-definition is interlaced, while enhanced-definition is progressive.

High-Definition

High-definition video is usually defined as having 720 progressive (720p) or 1080 interlaced (1080i) active scan lines. For a fixed-pixel (non-CRT) consumer display with a 16:9 aspect ratio, this translates into an active resolution of 1280 × 720p or 1920 × 1080i, respectively.

However, HDTV displays are technically defined as being capable of displaying a minimum of 720p or 1080i active scan lines. They also must be capable of displaying 16:9 content using a minimum of 540 progressive (540p) or 810 interlaced (810i) active scan lines. This enables the manufacturing of CRT-based HDTVs with a 4:3 aspect ratio and LCD/plasma 16:9 aspect ratio displays with resolutions of 1024 × 1024p, 1280 × 768p, 1024 × 768p, and so on, lowering costs.

VIDEO AND AUDIO COMPRESSION

The recent advances in consumer electronics, such as digital television, DVD players and recorders, digital video recorders, and so on, were made possible due to audio and video compression based largely on MPEG-2 video with Dolby® Digital, DTS®, MPEG-1, or MPEG-2 audio.

MPEG-2 expanded on the original MPEG-1 video and audio compression to cover a much wider range of applications. The primary application targeted during the definition process was all-digital transmission of broadcast-quality video at bit-rates of 4 to 9 Mbps. However, MPEG-2 is useful for many other applications, such as HDTV, and now supports bit-rates of 1.5 to 60 Mbps.

MPEG-2 is an ISO standard that consists of eleven parts. As with MPEG-1, the compressed bitstreams implicitly define the decompression algorithms. The compression algorithms are up to the individual manufacturers, within the scope of the international standard.

Today, in addition to the legacy DV, MPEG-1, and MPEG-2 audio and video compression standards, there are three new high-performance video compression standards, which offer much higher video compression for a given level of video quality:

- MPEG-4.2 This video codec typically offers a 1.5 to 2 times improvement in compression ratio over MPEG-2. However, this standard has never achieved widespread acceptance due to its complexity. Many simply decided to wait for the new MPEG-4.10 (H.264) video codec to become available.
- MPEG-4.10 (H.264) This video codec typically offers a two to three times improvement in compression ratio over MPEG-2. Additional improvements in compression ratios and quality are expected as the encoder become better and use more of the available tools that H.264 offers. H.264 is optimized for implementing on low-cost single-chip solutions and has already been adopted by the DVB and ARIB.
- SMPTE 421M (VC-1) A competitor to MPEG-4.10 (H.264), this video codec also typically offers a two to three times improvement in compression ratios over MPEG-2. Again, additional improvements in compression ratios and quality are expected as the encoders become better.

OTHER VIDEO APPLICATIONS

DVD Players

In addition to playing DVDs (which are based on MPEG-2 video compression), DVD players are expected to handle MP3 and WMA audio, MPEG-4 video (for DivX Video), JPEG images, and so on. Special playback modes such as slow/fast forward/reverse at various speeds are also expected. Support for DVD Audio and SACD is also popular.

A recent enhancement to DVD players is the ability to connect to a home network for playing content (music, video, pictures, etc.) residing on the PC. These "networked DVD players" may also include the ability to play movies from the Internet and download content onto an internal hard disc drive (HDD) for later viewing. Support for playing audio, video, and pictures from a variety of flash-memory cards is also growing.

How It Works

Looking at a simplified block diagram helps envision how video flows through its various operations. Figure 1.4 is a simplified block diagram for a basic DVD player, showing the common blocks. Today, all of this is on a single low-cost chip.

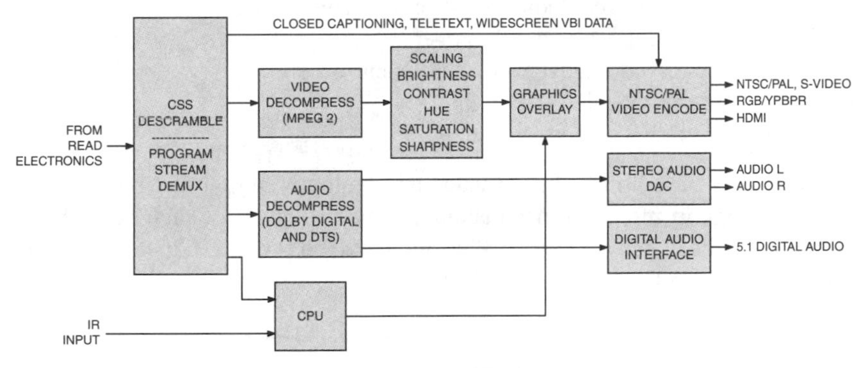

FIGURE 1.4 Simplified block diagram of a basic DVD player.

Digital Media Adapters

Digital media adapters connect to a home network for playing content (music, video, pictures, and so on) residing on a PC or media server. These small, low-cost boxes enable content to be easily enjoyed on any or all televisions in the home. Many support optional wireless networking, simplifying installation.

How It Works

Figure 1.5 is a simplified block diagram for a basic digital media adapter, showing the common blocks. Today, all of this is on a single low-cost chip.

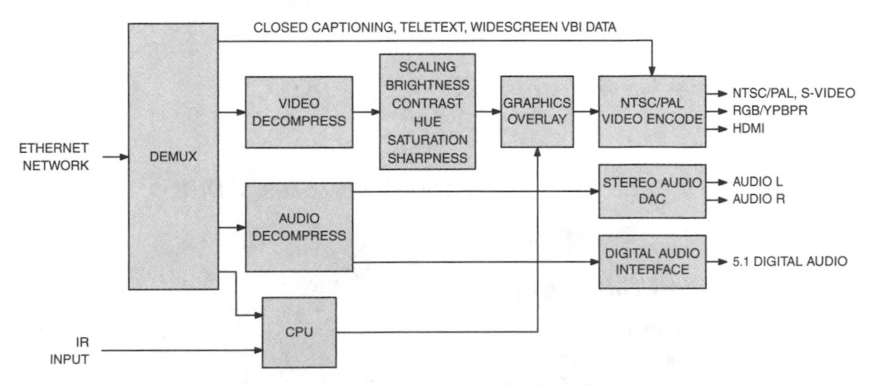

FIGURE 1.5 Simplified block diagram of a digital media adapter.

Digital Television Set-Top Boxes

The digital television standards fall into seven major categories:

1. ATSC (Advanced Television Systems Committee
2. DVB (Digital Video Broadcast)
3. ARIB (Association of Radio Industries and Businesses)
4. IPTV (including DVB and ARIB over IP)
5. Open digital cable standards, such as OpenCable
6. Proprietary digital cable standards
7. Proprietary digital satellite standards

Originally based on MPEG-2 video and Dolby® Digital or MPEG audio, they now support more advanced audio and video standards, such as MPEG-4 HE-AAC audio, Dolby® Digital Plus audio, MPEG-4.10 (H.264) video, and SMPTE 421 M (VC-1) video.

How It Works

Figure 1.6 is a simplified block diagram for a digital television set-top box, showing the common audio and video processing blocks. It is used to receive digital television broadcasts, from either terrestrial (over-the-air), cable, or satellite. A digital television may include this circuitry inside the television.

Many set-top boxes now include two tuners and digital video recorder (DVR) capability. This enables recording one program onto an internal HDD while watching another. Two tuners are also common in digital television receivers to support a picture-in-picture (PIP) feature.

STANDARDS ORGANIZATIONS

Many standards organizations, some of which are listed below, are involved in specifying video standards.

Advanced Television Systems Committee (ATSC)

(www.atsc.org)

Association of Radio Industries and Businesses (ARIB)

(www.arib.or.jp)

Cable Television Laboratories

(www.cablelabs.com)

Consumer Electronics Associations (CEA)

(www.ce.org)

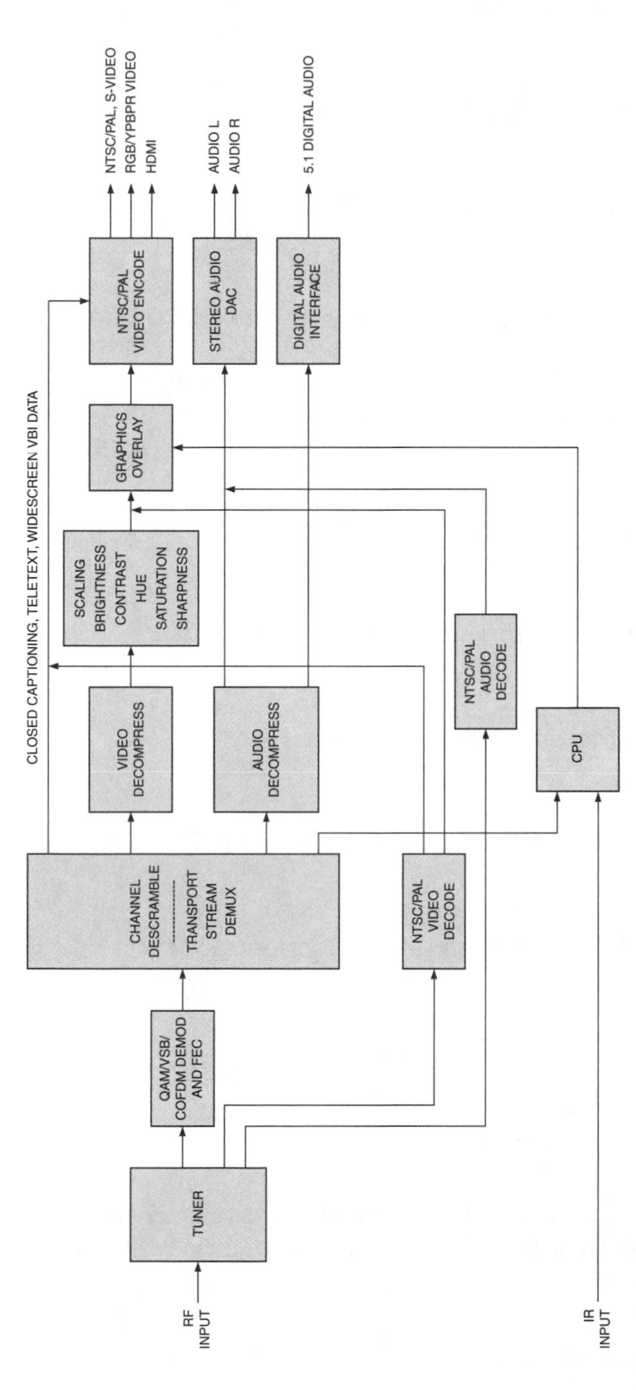

FIGURE 1.6 Simplified block diagram of a digital television set-top box.

Digital Video Broadcasting (DVB)
(www.dvb.org)

Electronic Industries Alliance (EIA)
(www.eia.org)

European Broadcasting Union (EBU)
(www.ebu.ch)

European Telecommunications Standards Institute (ETSI)
(www.etsi.org)

International Electrotechnical Commission (IEC)
(www.iec.ch)

Institute of Electrical and Electronics Engineers (IEEE)
(www.ieee.org)

International Organization for Standardization (ISO)
(www.iso.org)

International Telecommunication Union (ITU)
(www.itu.int)

Society of Cable Telecommunications Engineers (SCTE)
(www.scte.org)

Society of Motion Picture and Television Engineers (SMPTE)
(www.smpte.org)

Video Electronics Standards Association (VESA)
(www.vesa.org)

INSTANT SUMMARY

Implementing video on today's emerging products is a complex and challenging problem. There are many different facets and requirements to master. These include:

- Color spaces
- Digital and analog variations
- Connections

- Timing
- Resolution
  - Standard-definition
  - Enhanced-definition
  - High-definition
- Compression algorithms

Chapter 2

Color Spaces

In an Instant

- RGB Color Space
- YUV Color Space
- YIQ Color Space
- YCbCr Color Space
- HSI, HLS, and HSV Color Space
- Chromaticity Diagram
- Non-RGB Color Spaces
- Gamma Correction

Definitions

A color space is a mathematical representation of a set of colors. The three most popular color models are RGB (used in computer graphics); YIQ, YUV, or YCbCr (used in video systems); and CMYK (used in color printing). However, none of these color spaces is directly related to the intuitive notions of hue, saturation, and brightness. This resulted in the temporary pursuit of other models, such as HSI and HSV, to simplify programming, processing, and end-user manipulation.

All of the color spaces can be derived from the RGB information supplied by devices such as cameras and scanners.

RGB COLOR SPACE

The red, green, and blue (RGB) color space is widely used for computer graphics and displays. Red, green, and blue are three primary additive colors (meaning that individual components are added together to form a desired color) and are represented by a three-dimensional, Cartesian coordinate system (Figure 2.1). The indicated diagonal of the cube, with equal amounts of each primary component, represents various gray levels. Table 2.1 contains the RGB values for 100% amplitude, 100% saturated color bars, a common video test signal.

Technology Trade-offs

The RGB color space is the most prevalent choice for computer graphics because color displays use red, green, and blue to create the desired color. Therefore, the choice of the RGB color space simplifies the architecture and

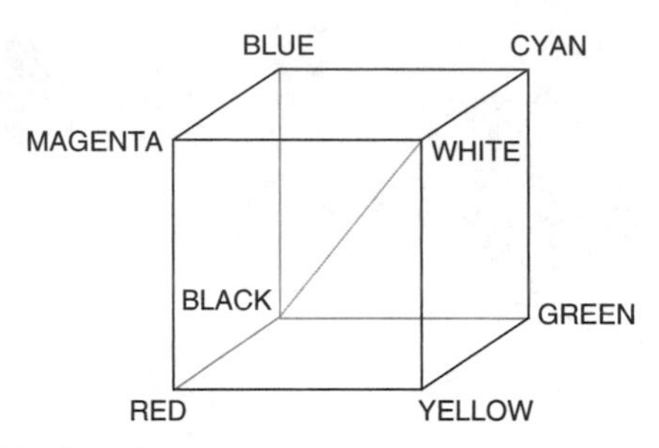

FIGURE 2.1 The RGB color cube.

TABLE 2.1 100% RGB color bars

	Nominal Range	White	Yellow	Cyan	Green	Magenta	Red	Blue	Black
R	0 to 255	255	255	0	0	255	255	0	0
G	0 to 255	255	255	255	255	0	0	0	0
B	0 to 255	255	0	255	0	255	0	255	0

design of the system. Also, a system that is designed using the RGB color space can take advantage of a large number of existing software routines, since this color space has been around for a number of years.

However, RGB is not very efficient when dealing with real-world images. All three RGB components need to be of equal bandwidth to generate any color within the RGB color cube. The result of this is a frame buffer that has the same pixel depth and display resolution for each RGB component. Also, processing an image in the RGB color space is usually not the most efficient method. For example, to modify the intensity or color of a given pixel, the three RGB values must be read from the frame buffer, the intensity or color calculated, the desired modifications performed, and the new RGB values calculated and written back to the frame buffer. If the system had access to an image stored directly in the intensity and color format, some processing steps would be faster.

For these and other reasons, many video standards use luma and two color difference signals. The most common are the YUV, YIQ, and YCbCr color spaces. Although all are related, there are some differences.

YUV COLOR SPACE

The YUV color space is used by the PAL (Phase Alternation Line), NTSC (National Television System Committee), and SECAM (Sequentiel Couleur

Avec Mémoire or Sequential Color with Memory) composite color video standards. The black-and-white system used only luma (Y) information; color information (U and V) was added in such a way that a black-and-white receiver would still display a normal black-and-white picture. Color receivers decoded the additional color information to display a color picture.

The basic equations to convert between gamma-corrected RGB (notated as R′G′B′—see the section titled Gamma Correction later in this chapter) and YUV are:

$$Y = 0.299R' + 0.587G' + 0.114B'$$

$$\begin{aligned} U &= -0.147R' - 0.289G' + 0.436B' \\ &= 0.492(B' - Y) \end{aligned}$$

$$\begin{aligned} V &= 0.615R' - 0.515G' - 0.100B' \\ &= 0.877(R' - Y) \end{aligned}$$

$$\begin{aligned} R' &= Y + 1.140V \\ G' &= Y - 0.395U - 0.581V \\ B' &= Y + 2.032U \end{aligned}$$

For digital R′G′B′ values with a range of 0–255, Y has a range of 0–255, U a range of 0 to ± 112, and V a range of 0 to ± 157. These equations are usually scaled to simplify the implementation in an actual NTSC or PAL digital encoder or decoder.

Note that for digital data, 8-bit YUV and R′G′B′ data should be saturated at the 0 and 255 levels to avoid underflow and overflow wrap-around problems.

Insider Info

If the full range of (B′ – Y) and (R′ – Y) had been used, the composite NTSC and PAL levels would have exceeded what the (then current) black-and-white television transmitters and receivers were capable of supporting. Experimentation determined that modulated subcarrier excursions of 20% of the luma (Y) signal excursion could be permitted above white and below black. The scaling factors were then selected so that the maximum level of 75% amplitude, 100% saturation yellow and cyan color bars would be at the white level (100 IRE).

YIQ COLOR SPACE

The YIQ color space is derived from the YUV color space and is optionally used by the NTSC composite color video standard. (The “I” stands for “in-phase” and the “Q” for “quadrature,” which is the modulation method used to transmit the color information.) The basic equations to convert between R′G′B′ and YIQ are:

$$Y = 0.299R' + 0.587G' + 0.114B'$$

$$\begin{aligned} I &= 0.596R' - 0.275G' - 0.321B' \\ &= V \cos 33° - U \sin 33° \\ &= 0.736(R' - Y) - 0.268(B' - Y) \end{aligned}$$

$$\begin{aligned} Q &= 0.212R' - 0.523G' + 0.31B' \\ &= V \sin 33° + U \cos 33° \\ &= 0.478(R' - Y) + 0.413(B' - Y) \end{aligned}$$

or, using matrix notation:

$$\begin{bmatrix} I \\ Q \end{bmatrix} = \begin{bmatrix} 0 & 1 \\ 1 & 0 \end{bmatrix} \begin{bmatrix} \cos(33) & \sin(33) \\ \sin(33) & \cos(33) \end{bmatrix} \begin{bmatrix} U \\ V \end{bmatrix}$$

$$\begin{aligned} R' &= Y + 0.956\,I + 0.621Q \\ G' &= Y - 0.272\,I - 0.647\,Q \\ B' &= Y - 1.107\,I + 1.704\,Q \end{aligned}$$

For digital R′G′B′ values with a range of 0–255, Y has a range of 0–255, I has a range of 0 to ±152, and Q has a range of 0 to ±134. I and Q are obtained by rotating the U and V axes 33°. These equations are usually scaled to simplify the implementation in an actual NTSC digital encoder or decoder.

Note that for digital data, 8-bit YIQ and R′G′B′ data should be saturated at the 0 and 255 levels to avoid underflow and overflow wrap-around problems.

YCBCR COLOR SPACE

The YCbCr color space was developed as part of ITU-R BT.601 during the development of a world-wide digital component video standard. YCbCr is a scaled and offset version of the YUV color space, and it is now used for all digital component video formats. Y is the luma component and the Cb and Cr are color difference signals. Y is defined to have a nominal 8-bit range of 16–235; Cb and Cr are defined to have a nominal range of 16–240. There are several YCbCr sampling formats, such as 4:4:4, 4:2:2, 4:1:1, and 4:2:0, defined as follows:

4:4:4 YCbCr means that for every Y sample, there is one sample each of Cb and Cr.

4:2:2 YCbCr means that for every two horizontal Y samples, there is one sample each of Cb and Cr.

4:1:1 YCbCr means that for every four horizontal Y samples, there is one sample each of Cb and Cr.

t0020

TABLE 2.2 75% YCbCr color bars

	Nominal Range	White	Yellow	Cyan	Green	Magenta	Red	Blue	Black
					SDTV				
Y	16 to 235	180	162	131	112	84	65	35	16
Cb	16 to 240	128	44	156	72	184	100	212	128
Cr	16 to 240	128	142	44	58	198	212	114	128
					HDTV				
Y	16 to 235	180	168	145	133	63	51	28	16
Cb	16 to 240	128	44	147	63	193	109	212	128
Cr	16 to 240	128	136	44	52	204	212	120	128

4:2:0 YCbCr means that for every block of 2 × 2 Y samples, there is one sample each of Cb and Cr. There are three variations of 4:2:0 YCbCr, with the difference being the position of Cb and Cr sampling relative to Y.

Table 2.2 lists the YCbCr values for 75% amplitude, 100% saturated color bars, a common video test signal.

Insider Info

The technically correct notation for YCbCr is Y'Cb'Cr' since all three components are derived from R'G'B'. However, most people use the YCbCr notation rather than Y'Cb'Cr'.

HSI, HLS, AND HSV COLOR SPACES

The HSI (hue, saturation, intensity) and HSV (hue, saturation, value) color spaces were developed to be more "intuitive" in manipulating color and were designed to approximate the way humans perceive and interpret color. They were developed when colors had to be specified manually, and are rarely used now that users can select colors visually or specify Pantone colors. These color spaces are discussed for historic interest. HLS (hue, lightness, saturation) is similar to HSI; the term lightness is used rather than intensity.

The difference between HSI and HSV is the computation of the brightness component (I or V), which determines the distribution and dynamic range of

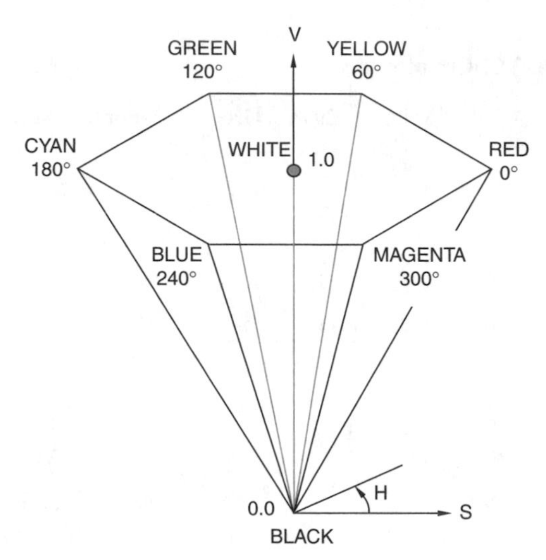

FIGURE 2.2 Single hexcone HSV color model.

both the brightness (I or V) and saturation (S). The HSI color space is best for traditional image processing functions such as convolution, equalization, histograms, and so on, which operate by manipulation of the brightness values since I is equally dependent on R, G, and B. The HSV color space is preferred for manipulation of hue and saturation (to shift colors or adjust the amount of color) since it yields a greater dynamic range of saturation.

Figure 2.2 illustrates the single hexcone HSV color model. The top of the hexcone corresponds to V = 1, or the maximum intensity colors. The point at the base of the hexcone is black and here V = 0. Complementary colors are 180° opposite one another as measured by H, the angle around the vertical axis (V), with red at 0°.The value of S is a ratio, ranging from 0 on the center line vertical axis (V) to 1 on the sides of the hexcone. Any value of S between 0 and 1 may be associated with the point V = 0. The point S = 0, V = 1 is white. Intermediate values of V for S = 0 are the grays. Note that when S = 0, the value of H is irrelevant. From an artist's viewpoint, any color with V = 1, S = 1 is a pure pigment (whose color is defined by H). Adding white corresponds to decreasing S (without changing V); adding black corresponds to decreasing V (without changing S). Tones are created by decreasing both S and V. Table 2.3 lists the 75% amplitude, 100% saturated HSV color bars.

Figure 2.3 illustrates the double hexcone HSI color model. The top of the hexcone corresponds to I = 1, or white. The point at the base of the hexcone is black and here I = 0. Complementary colors are 180° opposite one another as measured by H, the angle around the vertical axis (I), with red at 0° (for consistency with the HSV model, we have changed from the Tektronix convention

TABLE 2.3 75% HSV color bars

	Nominal Range	White	Yellow	Cyan	Green	Magenta	Red	Blue	Black
H	0° to 360°	–	60°	180°	120°	300°	0°	240°	–
S	0 to 1	0	1	1	1	1	1	1	0
V	0 to 1	0.75	0.75	0.75	0.75	0.75	0.75	0.75	0

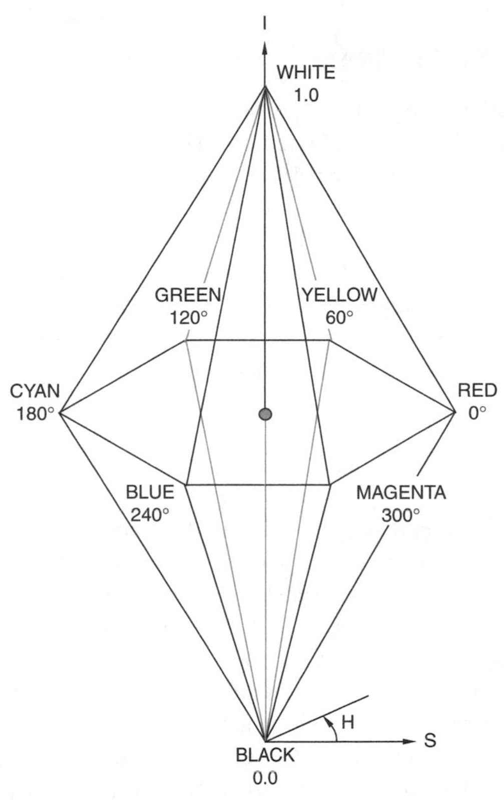

FIGURE 2.3 Double hexcone HIS color model. For consistency with the HSV model, we have changed from the Tektronix convention of blue at 0° and depict the model as a double hexcone rather than as a double cone.

of blue at 0°). The value of S ranges from 0 on the vertical axis (I) to 1 on the surfaces of the hexcone. The grays all have S = 0, but maximum saturation of hues is at S = 1, I = 0.5. Table 2.4 lists the 75% amplitude, 100% saturated HSI color bars.

TABLE 2.4 **75% HSI color bars. For consistency with the HSV model, we have changed from the Tektronix convention of blue at 0°**

	Nominal Range	White	Yellow	Cyan	Green	Magenta	Red	Blue	Black
H	0° to 360°	–	60°	180°	120°	300°	0°	240°	–
S	0 to 1	0	1	1	1	1	1	1	0
I	0 to 1	0.75	0.375	0.375	0.375	0.375	0.375	0.375	0

CHROMATICITY DIAGRAM

The color gamut perceived by a person with normal vision (the 1931 CIE Standard Observer) is shown in Figure 2.4. The diagram and underlying mathematics were updated in 1960 and 1976; however, the NTSC television system is based on the 1931 specifications.

Color perception was measured by viewing combinations of the three standard CIE (International Commission on Illumination or Commission Internationale de I'Eclairage) primary colors: red with a 700-nm wavelength, green at 546.1 nm, and blue at 435.8 nm.These primary colors, and the other spectrally pure colors resulting from mixing of the primary colors, are located along the curved outer boundary line (called the spectrum locus), shown in Figure 2.4.

The ends of the spectrum locus (at red and blue) are connected by a straight line that rep-resents the purples, which are combinations of red and blue. The area within this closed boundary contains all the colors that can be generated by mixing light of different colors. The closer a color is to the boundary, the more saturated it is. Colors within the boundary are perceived as becoming more pastel as the center of the diagram (white) is approached. Each point on the diagram, representing a unique color, may be identified by its x and y coordinates.

In the CIE system, the intensities of red, green, and blue are transformed into what are called the tristimulus values, which are represented by the capital letters X, Y, and Z. These values represent the relative quantities of the primary colors.

The coordinate axes of Figure 2.4 are derived from the tristimulus values:

$$x = X/(X + Y + Z) = \text{red}/(\text{red} + \text{green} + \text{blue})$$

$$y = Y/(X + Y + Z) = \text{green}/(\text{red} + \text{green} + \text{blue})$$

$$z = Z/(X + Y + Z) = \text{blue}/(\text{red} + \text{green} + \text{blue})$$

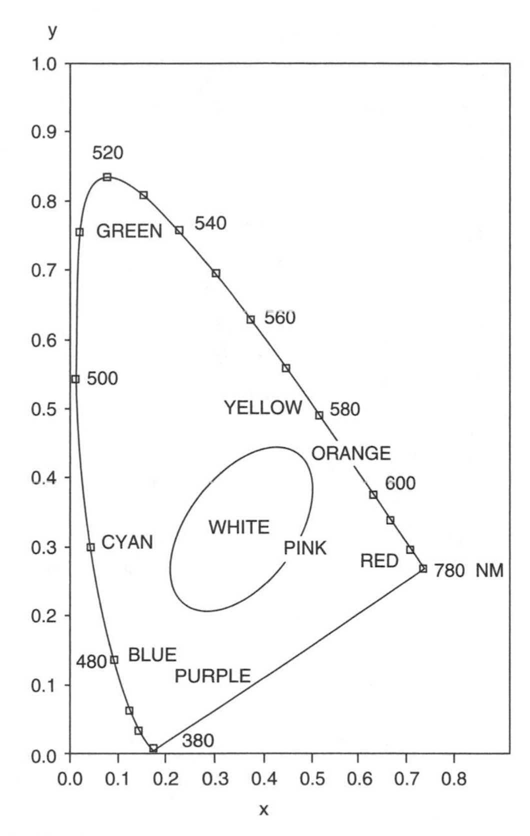

FIGURE 2.4 CIE 1931 chromaticity diagram showing various color regions.

The coordinates x, y, and z are called chromaticity coordinates, and they always add up to 1. As a result, z can always be expressed in terms of x and y, which means that only x and y are required to specify any color, and the diagram can be two-dimensional.

Typically, a source or display specifies three (x, y) coordinates to define the three primary colors it uses. The triangle formed by the three (x, y) coordinates encloses the gamut of colors that the source or display can reproduce. This is shown in Figure 2.5, which compares the color gamuts of NTSC, PAL and HDTV. Note that no set of three colors can generate all possible colors, which is why television pictures are never completely accurate.

In addition, a source or display usually specifies the (x, y) coordinate of the white color used, since pure white is not usually captured or reproduced. White is defined as the color captured or produced when all three primary signals are equal, and it has a subtle shade of color to it. Note that luminance, or brightness, information is not included in the standard CIE 1931 chromaticity diagram, but is an axis that is orthogonal to the (x, y) plane. The lighter a color is, the more restricted the chromaticity range is.

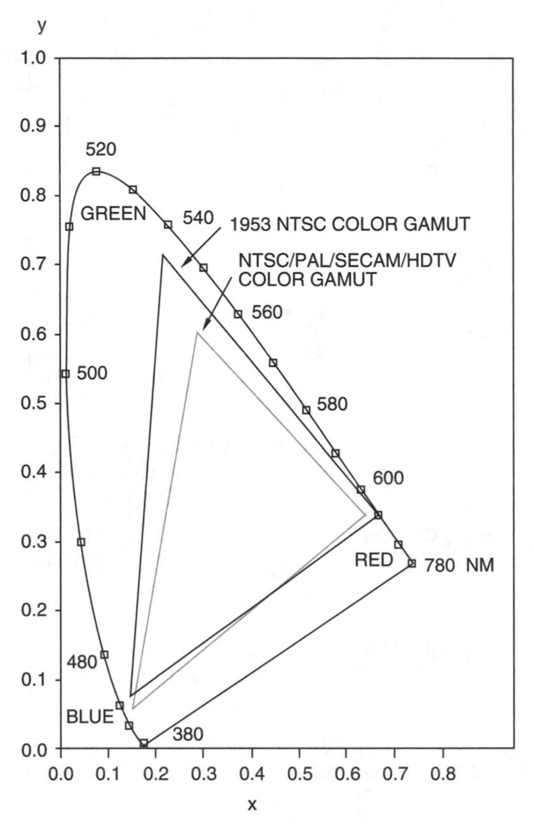

FIGURE 2.5 CIE 1931 chromaticity diagram showing various color gamuts.

The RGB chromaticities and reference white (CIE illuminate C) for the 1953 NTSC standard are:

$$
\begin{array}{rll}
R: & x_r = 0.67 & y_r = 0.33 \\
\text{G}: & x_g = 0.21 & y_g = 0.71 \\
B: & x_b = 0.14 & y_b = 0.08 \\
\textit{while}: & x_w = 0.3101 & y_w = 0.3162
\end{array}
$$

Modern NTSC, 480i and 480p video systems use a different set of RGB chromaticities (SMPTE "C") and reference white (CIE illuminate D65):

$$
\begin{array}{rll}
R: & x_r = 0.64 & y_r = 0.33 \\
\text{G}: & x_g = 0.29 & y_g = 0.60 \\
B: & x_b = 0.15 & y_b = 0.06 \\
\textit{while}: & x_w = 0.3127 & y_w = 0.3290
\end{array}
$$

The RGB chromaticities and reference white (CIE illuminate D65) for PAL, SECAM, 576i and 576p video systems are:

$$\begin{aligned} R&: \quad x_r = 0.64 \quad & y_r = 0.33 \\ G&: \quad x_g = 0.30 \quad & y_g = 0.60 \\ B&: \quad x_b = 0.15 \quad & y_b = 0.06 \\ \textit{while}&: \quad x_w = 0.3127 \quad & y_w = 0.3290 \end{aligned}$$

Since different chromaticity and reference white values are used for various video standards, minor color errors may occur when the source and display values do not match; for example, displaying a 480i or 480p program on an HDTV, or displaying an HDTV program on a NTSC television. These minor color errors can easily be corrected at the display by using a 3×3 matrix multiplier.

ALERT!

The RGB chromaticities for consumer displays are usually slightly different from the standards. As a result, one or more of the RGB colors are slightly off, such as having too much orange in the red, or too much blue in the green. This can usually be compensated by having the display professionally calibrated.

NON-RGB COLOR SPACE CONSIDERATIONS

When processing information in a non-RGB color space (such as YIQ, YUV, or YCbCr), care must be taken that combinations of values are not created that result in the generation of invalid RGB colors. The term "invalid" refers to RGB components outside the normalized RGB limits of (1, 1, 1).

For example, given that RGB has a normalized value of (1, 1, 1), the resulting YCbCr value is (235, 128, 128). If Cb and Cr are manipulated to generate a YCbCr value of (235, 64, 73), the corresponding RGB normalized value becomes (0.6, 1.29, 0.56)—note that the green value exceeds the normalized value of 1.

From this illustration it is obvious that there are many combinations of Y, Cb, and Cr that result in invalid RGB values; these YCbCr values must be processed so as to generate valid RGB values. Figure 2.6 shows the RGB normalized limits transformed into the YCbCr color space.

Best results are obtained using a constant luma and constant hue approach—Y is not altered while Cb and Cr are limited to the maximum valid values having the same hue as the invalid color prior to limiting. The constant hue principle corresponds to moving invalid CbCr combinations directly towards the CbCr origin (128, 128), until they lie on the surface of the valid YCbCr color block.

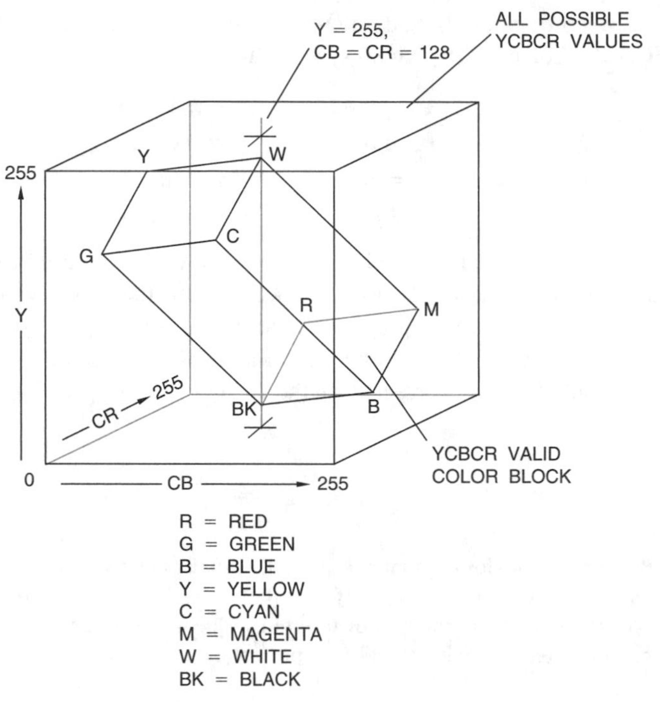

FIGURE 2.6 RGB limits transformed into 3-D YCbCr space.

ALERT!

When converting to the RGB color space from a non-RGB color space, care must be taken to include saturation logic to ensure overflow and underflow wrap-around conditions do not occur due to the finite precision of digital circuitry. 8-bit RGB values less than 0 must be set to 0, and values greater than 255 must be set to 255.

GAMMA CORRECTION

The transfer function of most CRT displays produces an intensity that is proportional to some power (referred to as gamma) of the signal amplitude. As a result, high-intensity ranges are expanded and low-intensity ranges are compressed (see Figure 2.7). This is an advantage in combatting noise, as the eye is approximately equally sensitive to equally relative intensity changes. By "gamma correcting" the video signals before transmission, the intensity output of the display is roughly linear (the gray line in Figure 2.7), and transmission-induced noise is reduced.

To minimize noise in the darker areas of the image, modern video systems limit the gain of the curve in the black region. This technique limits the gain

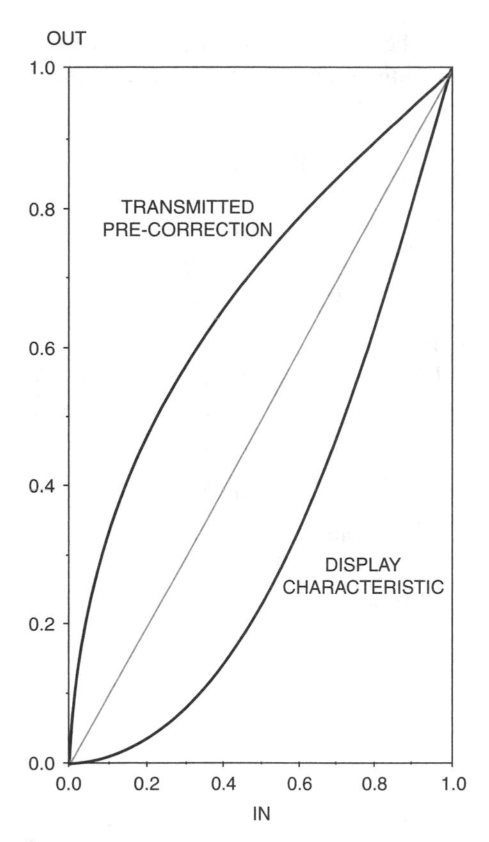

FIGURE 2.7 Effect of gamma.

close to black and stretches the remainder of the curve to maintain function and tangent continuity.

Technology Trade-offs

Although video standards assume a display gamma of about 2.2, a gamma of about 2.5 is more realistic for CRT displays. However, this difference improves the viewing in a dimly lit environment. More accurate viewing in a brightly lit environment may be accomplished by applying another gamma factor of about 1.14 (2.5/2.2). It is also common to tweak the gamma curve in the display to get closer to the "film look."

Early NTSC Systems

Early NTSC systems assumed a simple transform at the display, with a gamma of 2.2. RGB values are normalized to have a range of 0 to 1:

$$R = R'^{2.2}$$
$$G = G'^{2.2}$$
$$B = B'^{2.2}$$

To compensate for the nonlinear display, linear RGB data was "gamma-corrected" prior to transmission by the inverse transform. RGB values are normalized to have a range of 0 to 1:

$$R' = R^{1/2.2}$$
$$G' = G^{1/2.2}$$
$$B' = B^{1/2.2}$$

Early PAL and SECAM Systems

Most early PAL and SECAM systems assumed a simple transform at the display, with a gamma of 2.8. RGB values are normalized to have a range of 0 to 1:

$$R = R'^{2.8}$$
$$G = G'^{2.8}$$
$$B = B'^{2.8}$$

To compensate for the nonlinear display, linear RGB data was "gamma-corrected" prior to transmission by the inverse transform. RGB values are normalized to have a range of 0 to 1:

$$R' = R^{1/2.8}$$
$$G' = G^{1/2.8}$$
$$B' = B^{1/2.8}$$

Current Systems

Current NTSC, 480i, 480p, and HDTV video systems assume the following transform at the display, with a gamma of [1/0.45]. RGB values are normalized to have a range of 0 to 1:

if $(R', G', B') < 0.081$

$$R' = R'/4.5$$
$$G' = G'/4.5$$
$$B' = B'/4.5$$

if $(R', G', B') \geq 0.081$

$$R = ((R' + 0.099)/1.099)^{1/0.45}$$
$$G = ((G' + 0.099)/1.099)^{1/0.45}$$
$$B = ((B' + 0.099)/1.099)^{1/0.45}$$

To compensate for the nonlinear display, linear RGB data is "gamma-corrected" prior to transmission by the inverse transform. RGB values are normalized to have a range of 0 to 1:

if $(R, G, B) < 0.018$

$$R' = 4.5R$$
$$G' = 4.5G$$
$$B' = 4.5B$$

for (R, G, B) $\geq$0.018

$$R' = 1.099R^{0.45} - 0.099$$
$$G' = 1.099G^{0.45} - 0.099$$
$$B' = 1.099B^{0.45} - 0.099$$

Although most PAL and SECAM standards specify a gamma of 2.8, a value of [1/0.45] is now commonly used. Thus, these equations are also now used for PAL, SECAM, 576i, and 576p video systems.

Non-CRT Displays

Since they are not based on CRTs, the LCD, LCOS, DLP, and plasma displays have different display transforms. To simplify interfacing to these displays, their electronics are designed to accept standard gamma-corrected video and then compensate for the actual transform of the display panel.

INSTANT SUMMARY

Color spaces are mathematical representations of a set of colors and many different models are and have been used in video technology. These include:

- RGB color space
- YUV color space
- YIQ color space
- YCbCr color space
- HIS, HLS, and HSV color spaces

Chapter 3

Video Signals

In an Instant

- Video signal definitions
- Digital component video background
- 480i and 480p systems
- 576i and 576p systems
- 720p systems
- 1080i and 1080p systems

Definitions

Video signals come with a great variety of options—number of scan lines, interlaced vs. progressive, analog vs. digital, and so on. This chapter provides an overview of the common video signal formats and their timing. First, we will cover some definitions of terms that will be used in this chapter.

NTSC (National Television System Committee) is the name given to the color television transmission standard used in the U.S. Technically, NTSC is just a color modulation scheme; to specify the color video signal fully, it should be referred to as (M) NTSC. *PAL* (Phase Alternation Line) is the color television standard used in much of Europe, including the UK and Germany. Technically, PAL is just a color modulation scheme, and to fully specify the color video signal, it should be referred to as (B, D, G, H, I, M, N, or N_C) PAL. *SECAM* (Sequentiel Couleur Avec Memoire, or Sequential Color with Memory) is the color television transmission standard used in France and some other countries. It is similar to PAL but with some differences.

SDTV (Standard Definition Television) is content or a display that is capable of displaying a maximum of 576 interlaced active scan lines; no aspect ratio is specified for SDTV.

HDTV (High-Definition Television) is capable of displaying at 720 progressive or 1080 interlaced active scan lines. It must be capable of displaying a 16:9 image using at least 540 progressive or 810 interlaced active scan lines.

Active video is the part of the video signal that contains the picture information. Most of the active video, if not all of it, is visible on the display.

Aspect ratio is the ratio of the width of the picture to the height. Displays commonly have a 4:3 or 16:9 aspect ratio. Program material may have other aspect ratios (such as 2.35:1), resulting in its being letterboxed on the display.

Color bars are standard video test patterns used to check whether a video system is calibrated correctly.

On a CRT display, the scan line moves from the left edge to the right edge, jumps back to the left edge, and starts all over again, on down the screen. When the scan line hits the right side and is about to be brought back to the left side, the video signal is *blanked* so that you can't "see" the return path of the scan beam from the right to the left-hand edge. To blank the video signal, the analog video level is brought down to the *blanking level*, which is normally below the black level. *Pedestal* is an offset used to separate the black level from the blanking level by a small amount. When a video system doesn't use a pedestal, the black and blanking levels are the same.

The *frame rate* of a video source is how fast a new still image is available. 576i and 480i display originally used 25 and 30 frames per second, respectively. Refresh rates of 50, 60, 100, and 120 frames per second are now common.

Square pixels refers to a video display system in which the ratio of active pixels per line on a display is the same as the display aspect ratio.

Insider Info

"NTSC" is commonly, but incorrectly, used to refer to any 525/59.94 or 525/60 video system. "PAL" is commonly, but incorrectly, used to refer to any 625/50 video system.

DIGITAL COMPONENT VIDEO BACKGROUND

Key Concept

Digital component video ***is digital video that uses three separate color components, such as R′G′B′ or YCbCr. In digital component video, the video signals are in digital form (YCbCr or R′G′B′), being encoded to composite NTSC, PAL, or SECAM only when it is necessary for broadcasting or recording purposes.***

The European Broadcasting Union (EBU) became interested in a standard for digital component video due to the difficulties of exchanging video material between the 576i PAL and SECAM systems. The format held the promise that the digital video signals would be identical whether sourced in a PAL or SECAM country, allowing subsequent encoding to the appropriate composite form for broadcasting. Consultations with the Society of Motion Picture and Television Engineers (SMPTE) resulted in the development of an approach to support international program exchange, including 480i systems.

A series of demonstrations was carried out to determine the quality and suitability for signal processing of various methods. From these investigations, the main parameters of the digital component coding, filtering, and timing were chosen and incorporated into the ITU-R BT.601 standard. BT.601 has since served as the starting point for other digital component video standards.

Coding Ranges

The selection of the coding ranges balanced the requirements of adequate capacity for signals beyond the normal range and minimizing quantizing distortion. Although the black level of a video signal is reasonably well defined, the white level can be subject to variations due to video signal and equipment tolerances. Noise, gain variations, and transients produced by filtering can produce signal levels outside the nominal ranges.

8 or 10 bits per sample are used for each of the YCbCr or R′G′B′ components. Although 8-bit coding introduces some quantizing distortion, it was originally felt that most video sources contained sufficient noise to mask most of the quantizing distortion. However, if the video source is virtually noise-free, the quantizing distortion is noticeable as contouring in areas where the signal brightness gradually changes. In addition, at least two additional bits of fractional YCbCr or R′G′B′ data were desirable to reduce rounding effects when transmitting between equipment in the studio editing environment. For these reasons, most pro video equipment uses 10-bit YCbCr or R′G′B′, allowing 2 bits of fractional YCbCr or R′G′B′ data to be maintained.

Initial proposals had equal coding ranges for all three YCbCr components. However, this was changed so that Y had a greater margin for overloads at the white levels, as white level limiting is more visible than black. Thus, the nominal 8-bit Y levels are 16–235, while the nominal 8-bit CbCr levels are 16–240 (with 128 corresponding to no color). Occasional excursions into the other levels are permissible, but never at the 0 and 255 levels.

For 8-bit systems, the values of 0x00 and 0xFF are reserved for timing information. For 10-bit systems, the values of 0x000−0x003 and 0x3FC−0x3FF are reserved for timing information, to maintain compatibility with 8-bit systems.

The YCbCr or R′G′B′ levels to generate 75% color bars were discussed in Chapter 2. Digital R′G′B′ signals are defined to have the same nominal levels as Y to provide processing margin and simplify the digital matrix conversions between R′G′B′ and YCbCr.

SDTV Sample Rate Selection

Line-locked sampling of analog R′G′B′ or YUV video signals is specified for SDTV. This technique produces a static orthogonal sampling grid in which samples on the current scan line fall directly beneath those on previous scan lines and fields. It ensures that there is always a constant number of samples per scan line, even if the timing of the line changes.

Another important feature is that the sampling is locked in phase so that one sample is coincident with the 50% amplitude point of the falling edge of analog horizontal sync (0 × 0). This ensures that different sources produce samples at nominally the same positions in the picture. Making this feature common simplifies conversion from one standard to another.

For 480i and 576i video systems, several Y sampling frequencies were initially examined, including four times Fsc. However, the four-times Fsc sampling rates did not support the requirement of simplifying international exchange of programs, so they were dropped in favor of a single common sampling rate. Because the lowest sample rate possible (while still supporting quality video) was a goal, a 12 MHz sample rate was preferred for a long time, but eventually was considered to be too close to the Nyquist limit, complicating the filtering requirements. When the frequencies between 12 MHz and 14.3 MHz were examined, it became evident that a 13.5 MHz sample rate for Y provided some commonality between 480i and 576i systems. Cb and Cr, being color difference signals, do not require the same bandwidth as the Y, so may be sampled at one-half the Y sample rate, or 6.75 MHz.

Insider Info

The "4:2:2" notation now commonly used, and discussed in Chapter 2, originally applied to NTSC and PAL video, implying that Y, U and V were sampled at 4×, 2×, and 2× the color subcarrier frequency, respectively. The "4:2:2" notation was then adapted to BT.601 digital component video, implying that the sampling frequencies of Y, Cb and Cr were 4×, 2×, and 2× 3.375 MHz, respectively. "4:2:2" now commonly means that the sample rate of Cb and Cr is one-half that of Y, regardless of the actual sample rates used.

With 13.5 MHz sampling, each scan line contains 858 samples (480i systems) or 864 samples (576i systems) and consists of a digital blanking interval followed by an active line period. Both the 480i and 576i systems use 720 samples during the active line period. Having a common number of samples for the active line period simplifies the design of multistandard equipment and standards conversion. With a sample rate of 6.75 MHz for Cb and Cr (4:2:2 sampling), each active line period contains 360 Cr samples and 360 Cb samples.

Technology Trade-offs

With analog systems, problems may arise with repeated processing, causing an extension of the blanking intervals and softening of the blanking edges.

Using 720 digital samples for the active line period accommodates the range of analog blanking tolerances of both the 480i and 576i systems. Therefore, repeated processing may be done without affecting the digital blanking interval. Blanking to define the analog picture width need only be done once, preferably at the display or upon conversion to analog video.

Initially, BT.601 supported only 480i and 576i systems with a 4:3 aspect ratio (720 × 480i and 720 × 576i active resolutions). Support for a 16:9 aspect ratio was then added (960 × 480i and 960 × 576i active resolutions) using an 18 MHz sample rate.

EDTV Sample Rate Selection

ITU BT.1358 defines the progressive SDTV video signals, also known as 480p or 576p, or Enhanced Digital Television (EDTV). The sample rate is doubled to 27 MHz (4:3 aspect ratio) or 36 MHz (16:9 aspect ratio) in order to keep the same static orthogonal sampling grid as that used by BT.601.

HDTV Sample Rate Selection

ITU BT.709 defines the 720p, 1080i, and 1080p video signals, respectively. With HDTV, a different technique was used—the number of active samples per line and the number of active lines per frame is constant, regardless of the frame rate. Thus, in order to keep a static orthogonal sampling grid, each frame rate uses a different sample clock rate.

480i AND 480p SYSTEMS

Interlaced Analog Composite Video

(M) NTSC and (M) PAL are analog composite video signals that carry all timing and color information within a single signal. These analog interfaces use 525 lines per frame.

Interlaced Analog Component Video

Analog component signals are comprised of three signals, analog R′G′B′ or YPbPr. Referred to as 480i (since there are typically 480 active scan lines per frame and they are interlaced), the frame rate is usually 29.97 Hz (30/1.001) for compatibility with (M) NTSC timing. The analog interface uses 525 lines per frame, with active video present on lines 23–262 and 286–525, as shown in Figure 3.1.

For the 29.97 Hz frame rate, each scan line time (H) is about 63.556 μs. Detailed horizontal timing is dependent on the specific video interface used, which will be covered in Chapter 4.

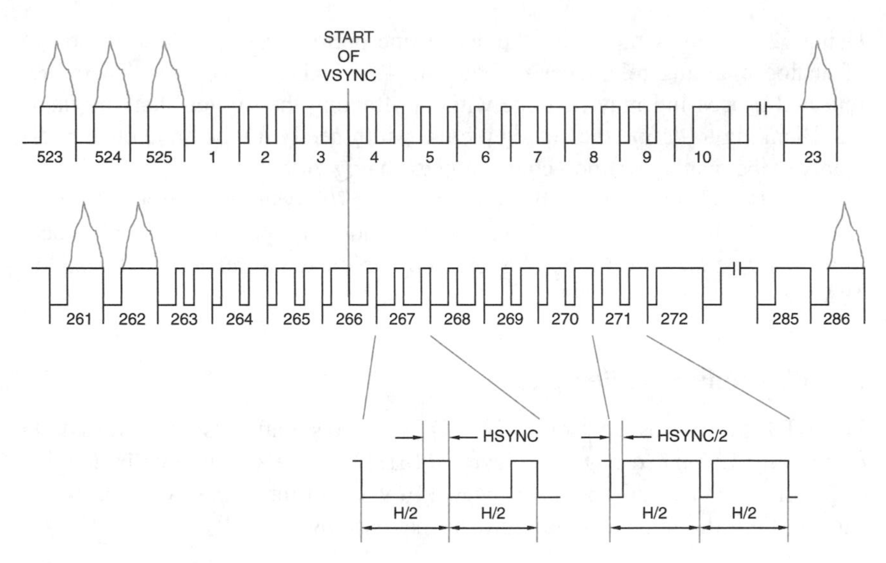

FIGURE 3.1 480i Vertical Interval Timing.

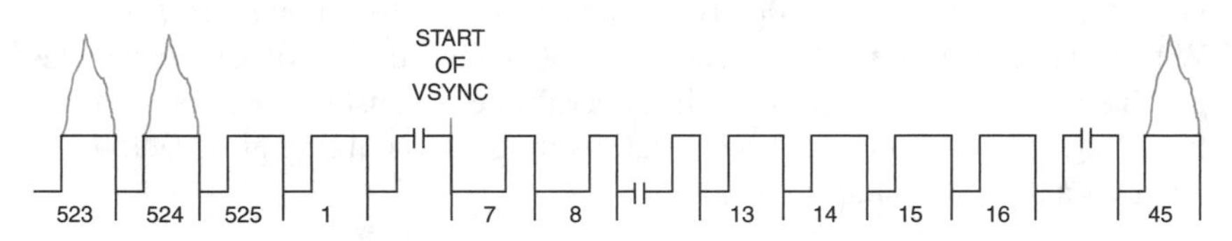

FIGURE 3.2 480p Vertical Interval Timing.

Progressive Analog Component Video

Analog component signals are comprised of three signals, analog R′G′B′ or YPbPr. Referred to as 480p (since there are typically 480 active scan lines per frame and they are progressive), the frame rate is usually 59.94 Hz (60/1.001) for easier compatibility with (M) NTSC timing. The analog interface uses 525 lines per frame, with active video present on lines 45–524, as shown in Figure 3.2.

For the 59.94 Hz frame rate, each scan line time (H) is about 31.776 μs. Detailed horizontal timing is dependent on the specific video interface used, as discussed in Chapter 4.

Interlaced Digital Component Video

BT.601 and SMPTE 267M specify the representation for 480i digital R′G′B′ or YCbCr video signals. Active resolutions defined within BT.601 and SMPTE

267M, along with their 1× Y and R′G′B′ sample rates (Fs), and frame rates, are as follows:

960 × 480i	18.0 MHz	29.97 Hz
720 × 480i	13.5 MHz	29.97 Hz

Other common active resolutions, their 1× sample rates (Fs), and frame rates, are:

864 × 480i	16.38 MHz	29.97 Hz
704 × 480i	13.50 MHz	29.97 Hz
640 × 480i	12.27 MHz	29.97 Hz
544 × 480i	10.12 MHz	29.97 Hz
528 × 480i	9.900 MHz	29.97 Hz
480 × 480i	9.900 MHz	29.97 Hz
352 × 480i	6.750 MHz	29.97 Hz

864 × 480i is a 16:9 square pixel format, while 640 × 480i is a 4:3 square pixel format. Although the ideal 16:9 resolution is 854 × 480i, 864 × 480i supports the MPEG 16 × 16 block structure. The 704 × 480i format is done by using the 720 × 480i format, and blanking the first eight and last eight samples in each active scan line. Example relationships between the analog and digital signals are shown in Figures 3.3 through 3.7.

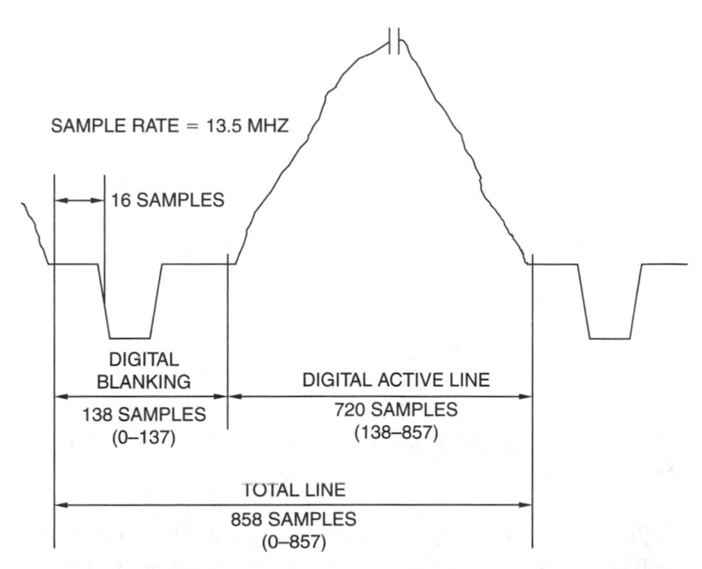

FIGURE 3.3 480i Analog-Digital Relationship (4:3 Aspect Ratio, 29.97 Hz Frame Rate, 13.5 MHz Sample Clock). BT.601 specifies 16 samples for the front porch; CEA-861D (DVI and HDMI timing) specifies 19 samples for the front porch.

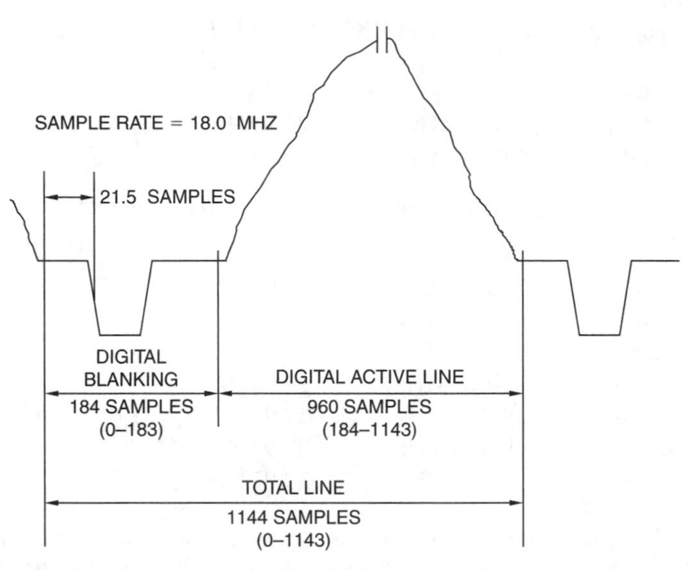

FIGURE 3.4 480i Analog-Digital Relationship (16:9 Aspect Ratio, 29.97 Hz Frame Rate, 18 MHz Sample Clock).

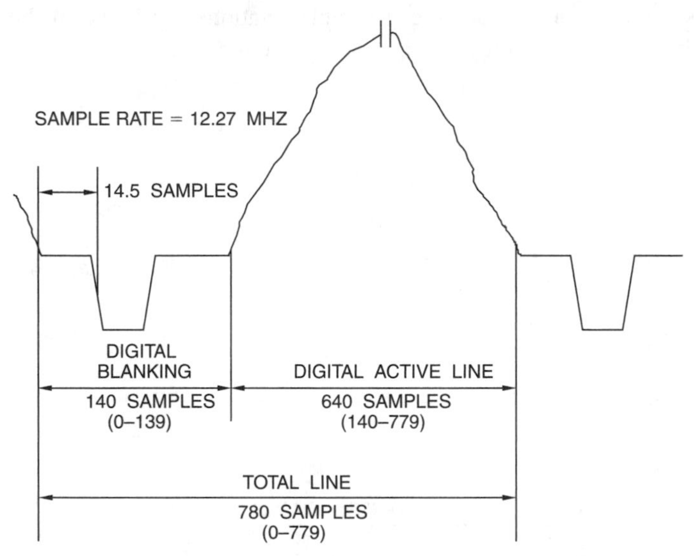

FIGURE 3.5 480i Analog-Digital Relationship (4:3 Aspect Ratio, 29.97 Hz Frame Rate, 12.27 MHz Sample Clock).

The H (horizontal blanking), V (vertical blanking), and F (field) signals are defined in Figure 3.8. The H, V, and F timing indicated is compatible with video compression standards rather than BT.656 discussed in Chapter 5.

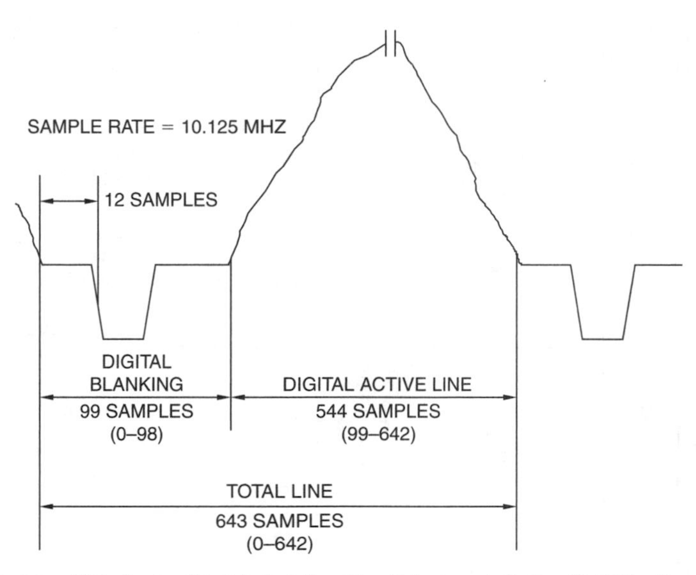

FIGURE 3.6 480i Analog-Digital Relationship (4:3 Aspect Ratio, 29.97 Hz Frame Rate, 10.125 MHz Sample Clock).

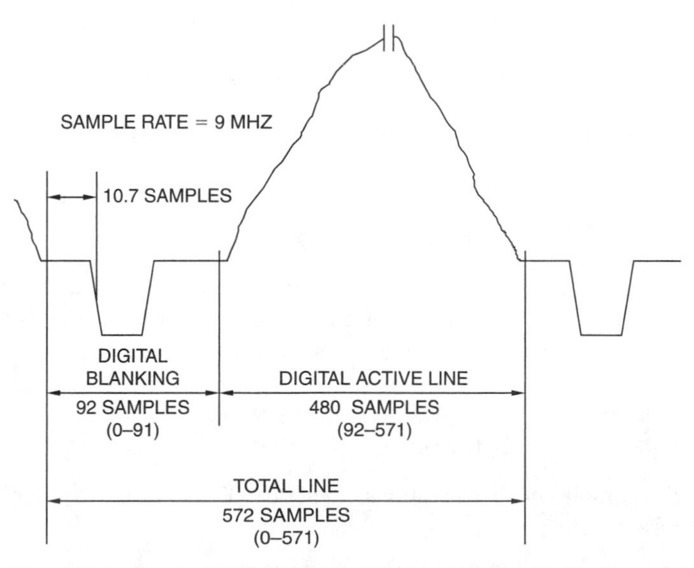

FIGURE 3.7 480i Analog-Digital Relationship (4:3 Aspect Ratio, 29.97 Hz Frame Rate, 9 MHz Sample Clock).

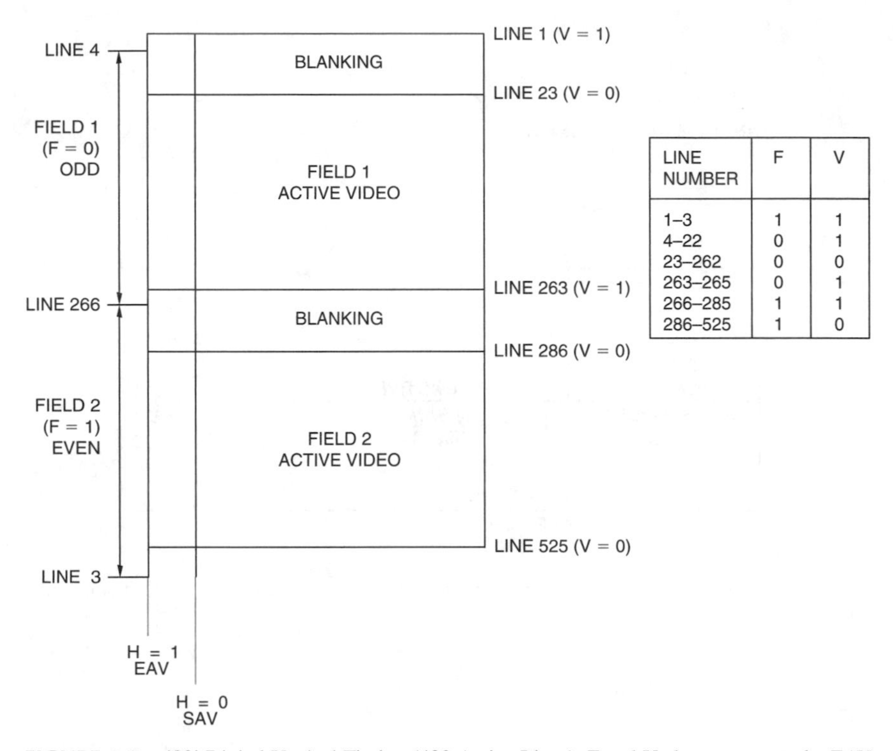

LINE NUMBER	F	V
1–3	1	1
4–22	0	1
23–262	0	0
263–265	0	1
266–285	1	1
286–525	1	0

FIGURE 3.8 480i Digital Vertical Timing (480 Active Lines). F and V change state at the EAV sequence at the beginning of the digital line. Note that the digital line number changes state prior to the start of horizontal sync, as shown in Figures 3.3 through 3.7.
These active lines are used by the SMPTE RP-202, ATSC A/54a, and ARIB STD-B32 standards. CEA-861D (DVI and HDMI timing) specifies lines 22–261 and 285–524 for active video. IEC 61834-2, ITU-R BT.1618, and SMPTE 314M (DV formats) specify lines 23–262 and 285–524 for active video.
ITU-R BT.656 specifies lines 20–263 and 283–525 for active video, resulting in 487 total active lines per frame.

Progressive Digital Component Video

BT.1358 and SMPTE 293M specify the representation for 480p digital R′G′B′ or YCbCr video signals. Active resolutions defined within BT.1358 and SMPTE 293M, along with their 1× sample rates (Fs), and frame rates, are:

960 × 480p	36.0 MHz	59.94 Hz
720 × 480p	27.0 MHz	59.94 Hz

Other common active resolutions, their 1× Y and R′G′B′ sample rates (Fs), and frame rates, are:

864 × 480p	32.75 MHz	59.94 Hz
704 × 480p	27.00 MHz	59.94 Hz

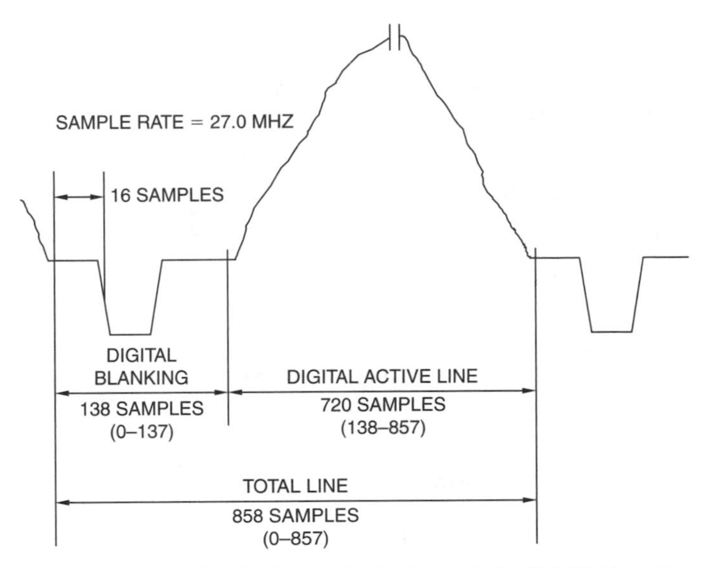

FIGURE 3.9 480p Analog-Digital Relationship (4:3 Aspect Ratio, 59.94 Hz Frame Rate, 27 MHz Sample Clock).

640 × 480p	24.54 MHz	59.94 Hz
544 × 480p	20.25 MHz	59.94 Hz
528 × 480p	19.80 MHz	59.94 Hz
480 × 480p	18.00 MHz	59.94 Hz
352 × 480p	13.50 MHz	59.94 Hz

864 × 480p is a 16:9 square pixel format, while 640 × 480p is a 4:3 square pixel format. Although the ideal 16:9 resolution is 854 × 480p, 864 × 480p supports the MPEG 16 × 16 block structure. The 704 × 480p format is done by using the 720 × 480p format, and blanking the first eight and last eight samples each active scan line. Example relationships between the analog and digital signals are shown in Figures 3.9 through 3.12.

The H (horizontal blanking), V (vertical blanking), and F (field) signals are defined in Figure 3.13. The H, V, and F timing indicated is compatible with video compression standards rather than BT.656 discussed in Chapter 5.

576i AND 576p SYSTEMS

Interlaced Analog Composite Video

(B, D, G, H, I, N, NC) PAL are analog composite video signals that carry all timing and color information within a single signal. These analog interfaces use 625 lines per frame.

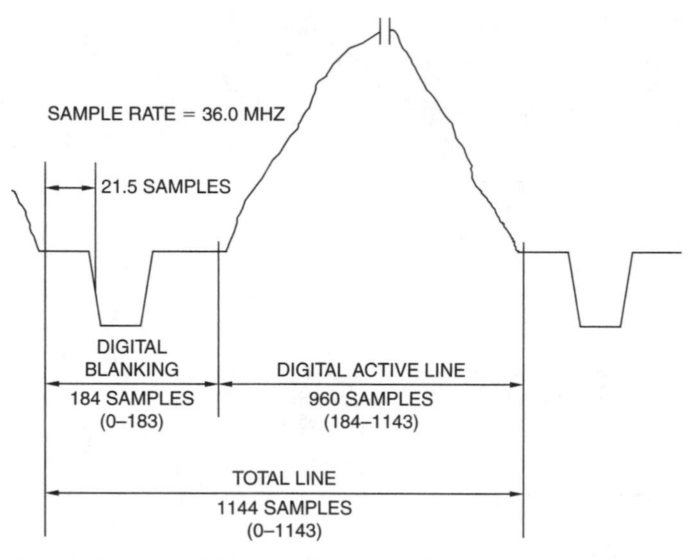

FIGURE 3.10 480p Analog-Digital Relationship (16:9 Aspect Ratio, 59.94 Hz Frame Rate, 36 MHz Sample Clock).

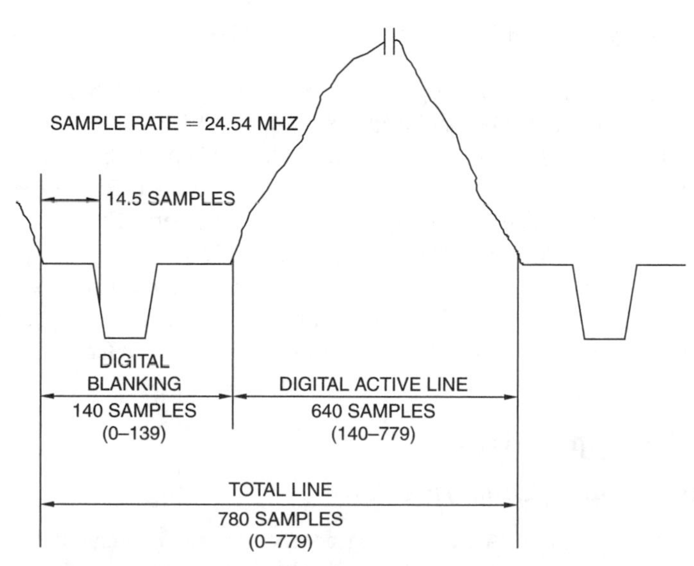

FIGURE 3.11 480p Analog-Digital Relationship (4:3 Aspect Ratio, 59.94 Hz Frame Rate, 24.54 MHz Sample Clock).

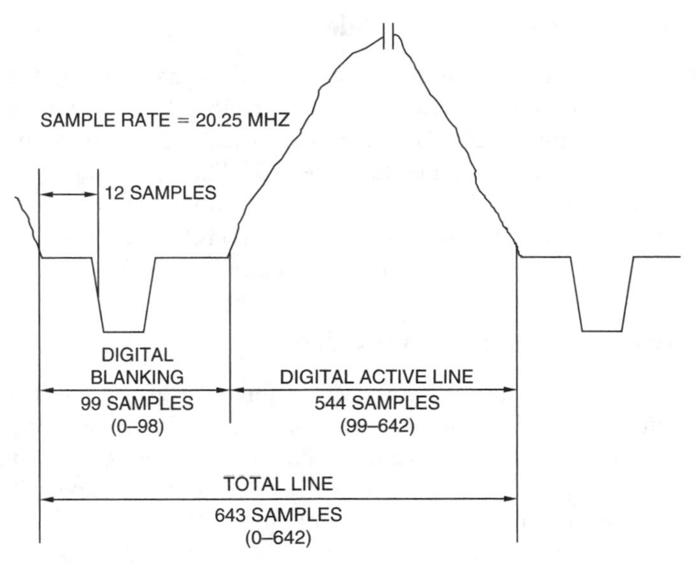

FIGURE 3.12 480p Analog-Digital Relationship (4:3 Aspect Ratio, 59.94 Hz Frame Rate, 20.25 MHz Sample Clock).

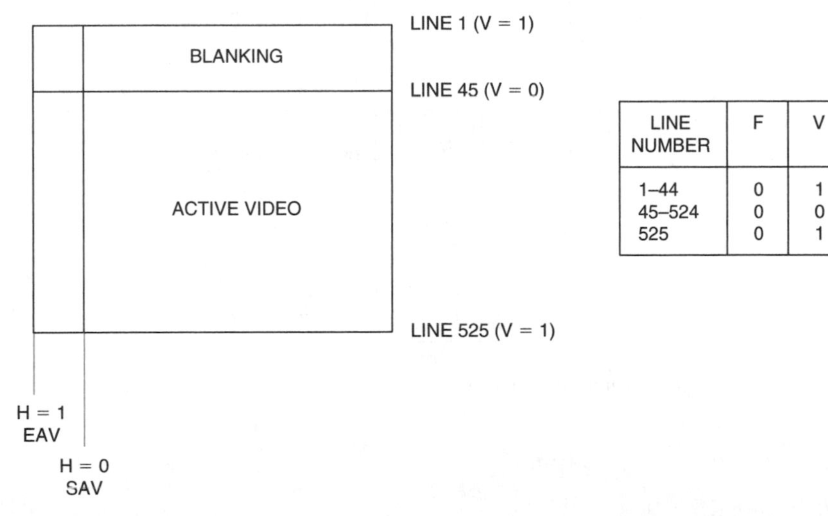

LINE NUMBER	F	V
1–44	0	1
45–524	0	0
525	0	1

FIGURE 3.13 480p Digital Vertical Timing (480 Active Lines). V changes state at the EAV sequence at the beginning of the digital line. Note that the digital line number changes state prior to the start of horizontal sync, as shown in Figures 3.9 through 3.12.
These active lines are used by the SMPTE RP-202, ATSC A/54, and ARIB STD-B32 standards. However, CEA-861 (DVI and HDMI timing) specifies lines 43–522 for active video.

Interlaced Analog Component Video

Analog component signals are comprised of three signals, analog R′G′B′ or YPbPr. Referred to as 576i (since there are typically 576 active scan lines per frame and they are interlaced), the frame rate is usually 25 Hz for compatibility with PAL timing. The analog interface uses 625 lines per frame, with active video present on lines 23–310 and 336–623, as shown in Figure 3.14.

For the 25 Hz frame rate, each scan line time (H) is 64 μs. Detailed horizontal timing is dependent on the specific video interface used, as discussed in Chapter 4.

Progressive Analog Component Video

Analog component signals are comprised of three signals, analog R′G′B′ or YPbPr. Referred to as 576p (since there are typically 576 active scan lines per frame and they are progressive), the frame rate is usually 50 Hz for compatibility with PAL timing. The analog interface uses 625 lines per frame, with active video present on lines 45–620, as shown in Figure 3.15.

For the 50 Hz frame rate, each scan line time (H) is 32 μs. Detailed horizontal timing is dependent on the specific video interface used, as discussed in Chapter 4.

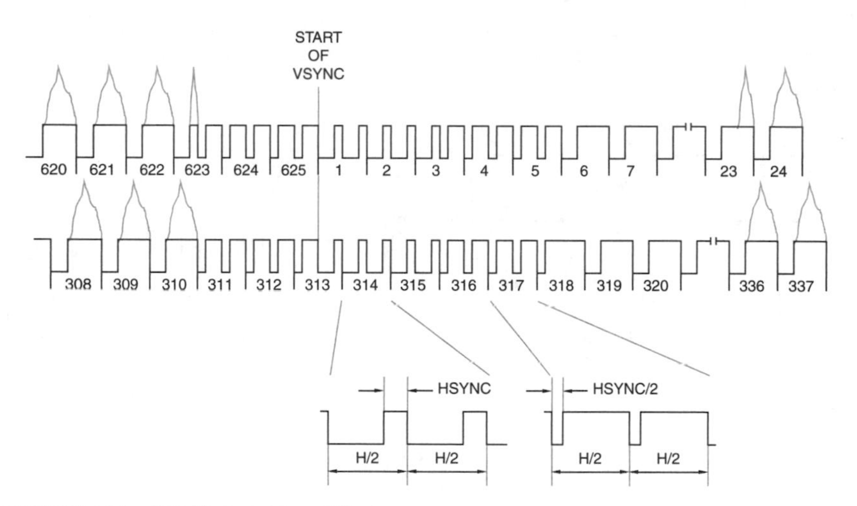

FIGURE 3.14 576i Vertical Interval Timing.

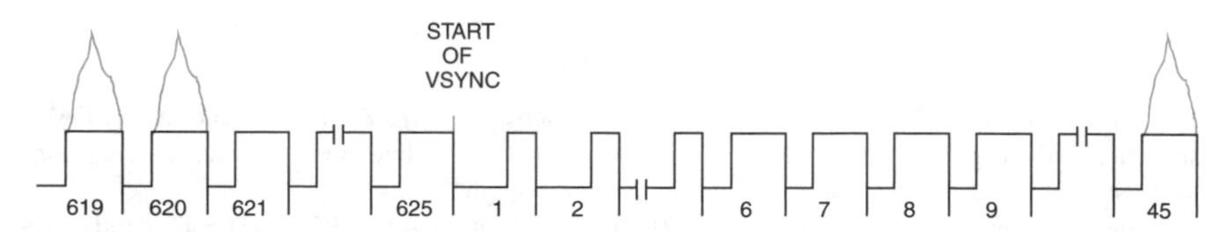

FIGURE 3.15 576p Vertical Interval Timing.

Interlaced Digital Component Video

BT.601 specifies the representation for 576i digital R′G′B′ or YCbCr video signals. Active resolutions defined within BT.601, their 1× Y and R′G′B′ sample rates (Fs), and frame rates, are:

960 × 576i	18.0 MHz	25 Hz
720 × 576i	13.5 MHz	25 Hz

Other common active resolutions, their 1× ?Y and R′G′B′ sample rates (Fs), and frame rates, are:

1024 × 576i	19.67 MHz	25 Hz
768 × 576i	14.75 MHz	25 Hz
704 × 576i	13.50 MHz	25 Hz
544 × 576i	10.12 MHz	25 Hz
480 × 576i	9.000 MHz	25 Hz

1024 × 576i is a 16:9 square pixel format, while 768 × 576i is a 4:3 square pixel format. The 704 × 576i format is done by using the 720 × 576i format, and blanking the first eight and last eight samples each active scan line. Example relationships between the analog and digital signals are shown in Figures 3.16 through 3.19.

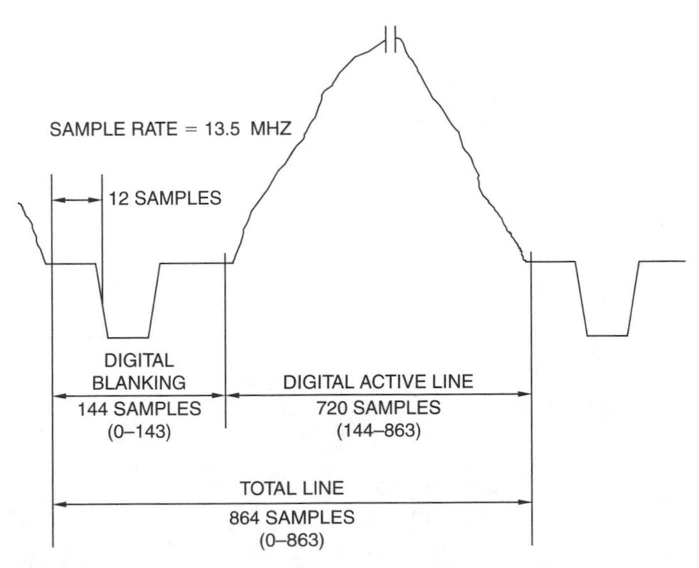

FIGURE 3.16 576i Analog-Digital Relationship (4:3 Aspect Ratio, 25 Hz Frame Rate, 13.5 MHz Sample Clock).

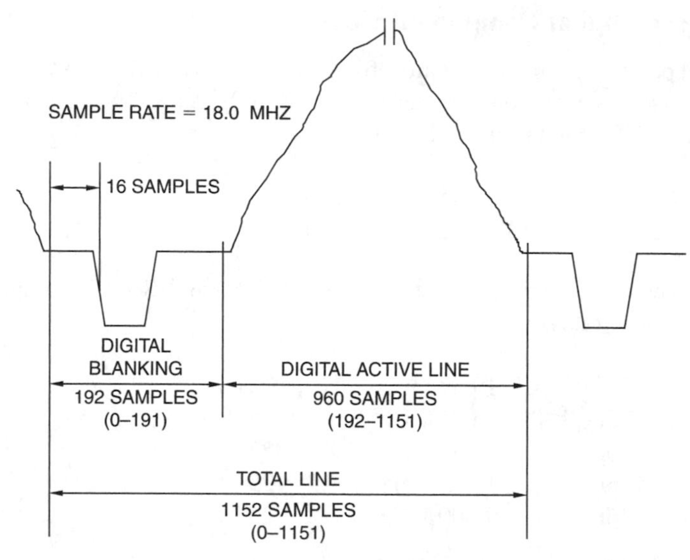

FIGURE 3.17 576i Analog-Digital Relationship (16:9 Aspect Ratio, 25 Hz Frame Rate, 18 MHz Sample Clock).

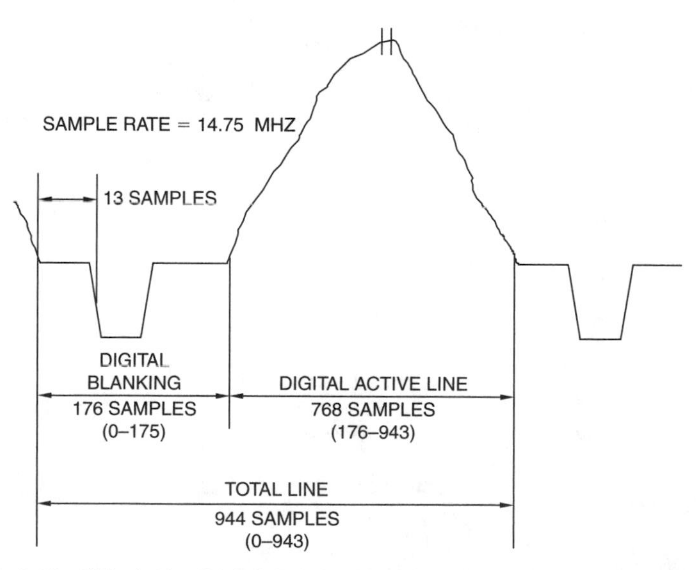

FIGURE 3.18 576i Analog-Digital Relationship (4:3 Aspect Ratio, 25 Hz Frame Rate, 14.75 MHz Sample Clock).

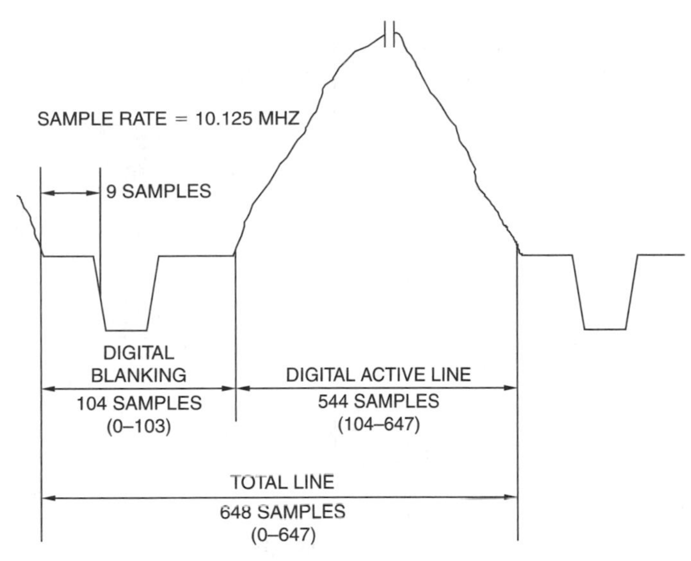

FIGURE 3.19 576i Analog-Digital Relationship (4:3 Aspect Ratio, 25 Hz Frame Rate, 10.125 MHz Sample Clock).

The H (horizontal blanking), V (vertical blanking), and F (field) signals are defined in Figure 3.20. The H, V, and F timing indicated is compatible with video compression standards rather than BT.656 discussed in Chapter 5.

Progressive Digital Component Video

BT.1358 specifies the representation for 576p digital R′G′B′ or YCbCr signals. Active resolutions defined within BT.1358, their 1× Y and R′G′B′ sample rates (Fs), and frame rates, are:

960 × 576p	36.0 MHz	50 Hz
720 × 576p	27.0 MHz	50 Hz

Other common active resolutions, their 1× Y and R′G′B′ sample rates (Fs), and frame rates, are:

1024 × 576p	39.33 MHz	50 Hz
768 × 576p	29.50 MHz	50 Hz
704 × 576p	27.00 MHz	50 Hz
544 × 576p	20.25 MHz	50 Hz
480 × 576p	18.00 MHz	50 Hz

1024 × 576p is a 16:9 square pixel format, while 768 × 576p is a 4:3 square pixel format. The 704 × 576p format is done by using the 720 × 576p

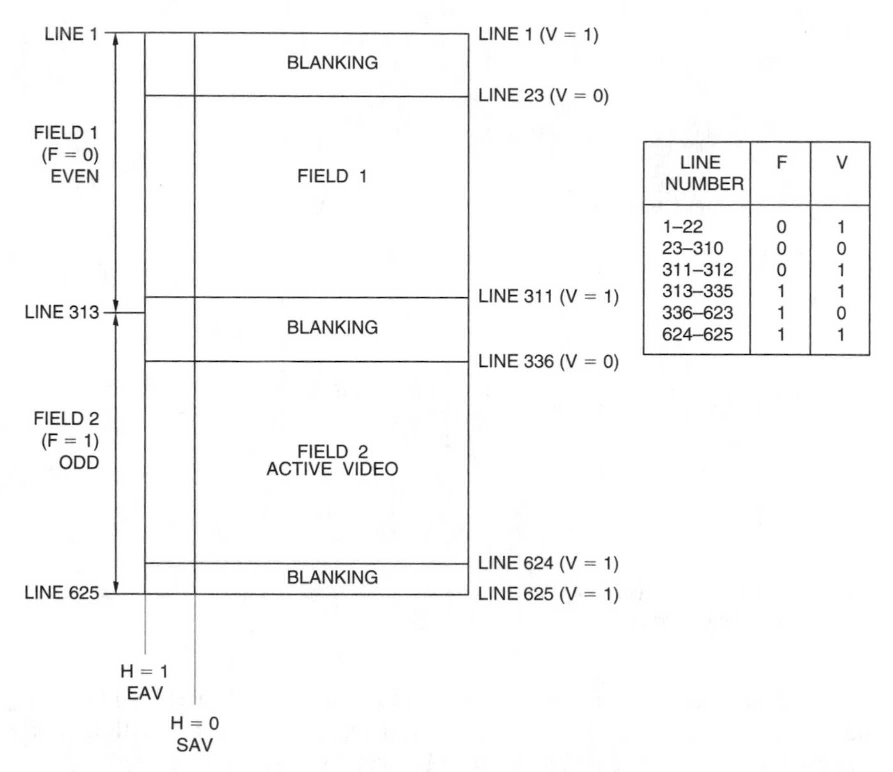

LINE NUMBER	F	V
1–22	0	1
23–310	0	0
311–312	0	1
313–335	1	1
336–623	1	0
624–625	1	1

FIGURE 3.20 576i Digital Vertical Timing (576 Active Lines). F and V change state at the EAV sequence at the beginning of the digital line. Note that the digital line number changes state prior to the start of horizontal sync, as shown in Figures 4.16 through 4.19. IEC 61834-2, ITU-R BT.1618, and SMPTE 314M (DV formats) specify lines 23–310 and 335–622 for active video.

format, and blanking the first eight and last eight samples each active scan line. Example relationships between the analog and digital signals are shown in Figures 3.21 through 3.24.

The H (horizontal blanking), V (vertical blanking), and F (field) signals are defined in Figure 3.25. The H, V, and F timing indicated is compatible with video compression standards rather than BT.656 discussed in Chapter 5.

720p SYSTEMS

Progressive Analog Component Video

Analog component signals are comprised of three signals, analog R′G′B′ or YPbPr. Referred to as 720p (since there are typically 720 active scan lines per frame and they are progressive), the frame rate is usually 59.94 Hz (60/1.001) to simplify the generation of (M) NTSC video. The analog interface uses 750 lines per frame, with active video present on lines 26–745, as shown in Figure 3.26.

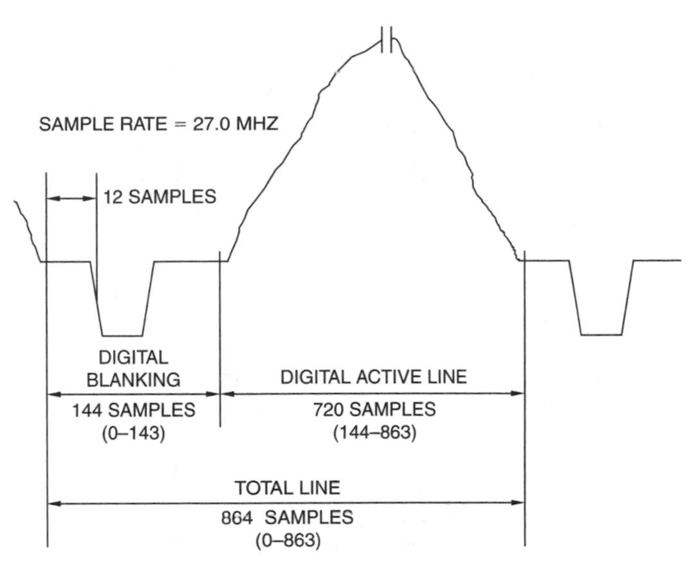

FIGURE 3.21 576p Analog-Digital Relationship (4:3 Aspect Ratio, 50 Hz Frame Rate, 27 MHz Sample Clock).

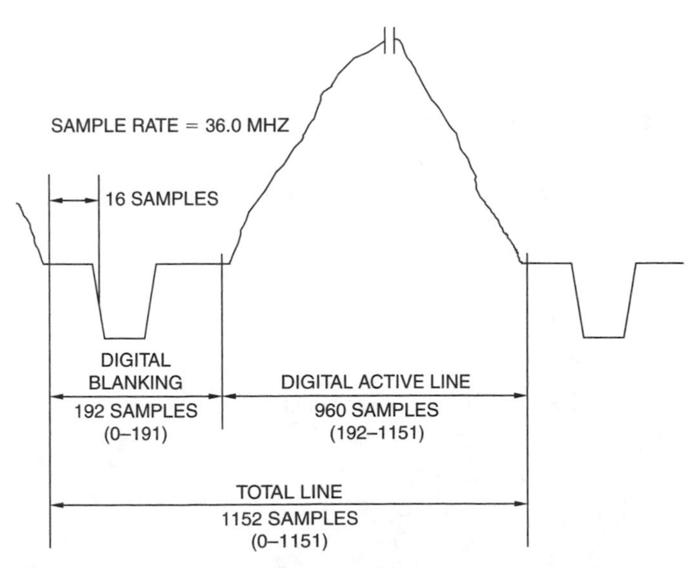

FIGURE 3.22 576p Analog-Digital Relationship (16:9 Aspect Ratio, 50 Hz Frame Rate, 36 MHz Sample Clock).

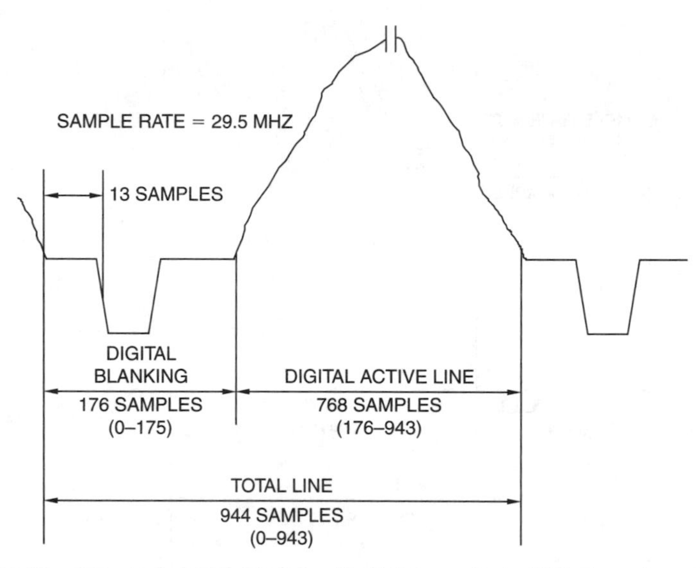

FIGURE 3.23 576p Analog-Digital Relationship (4:3 Aspect Ratio, 50 Hz Frame Rate, 29.5 MHz Sample Clock).

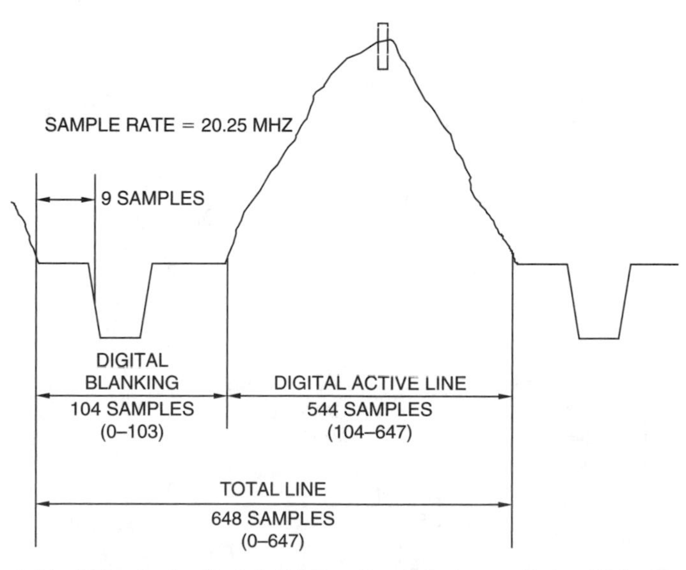

FIGURE 3.24 576p Analog-Digital Relationship (4:3 Aspect Ratio, 50 Hz Frame Rate, 20.25 MHz Sample Clock).

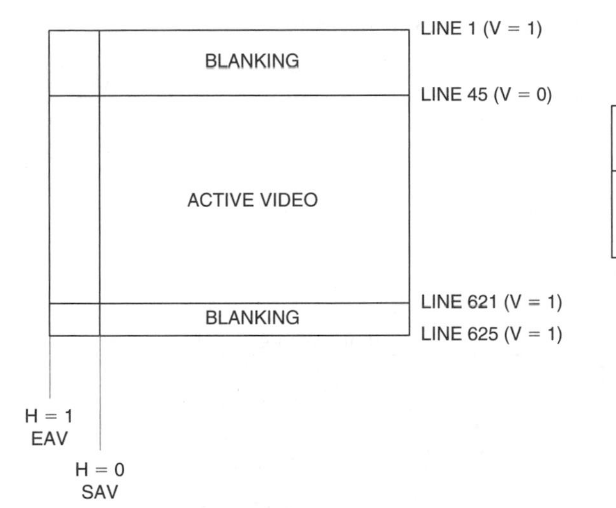

LINE NUMBER	F	V
1–44	0	1
45–620	0	0
621–625	0	1

FIGURE 3.25 576p Digital Vertical Timing (576 Active Lines). V changes state at the EAV sequence at the beginning of the digital line. Note that the digital line number changes state prior to the start of horizontal sync, as shown in Figures 3.21 through 3.24.

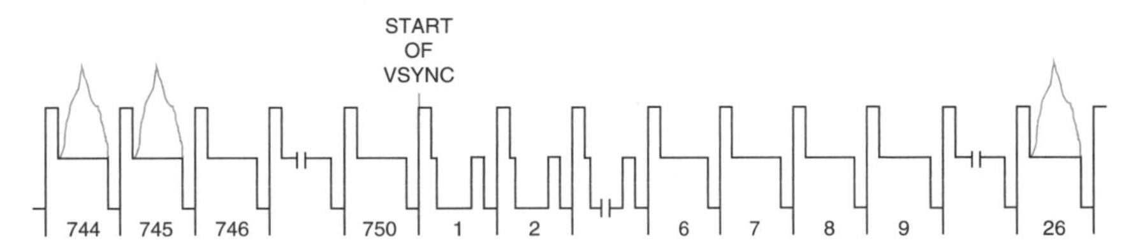

FIGURE 3.26 720p Vertical Interval Timing.

For the 59.94 Hz frame rate, each scan line time (H) is about 22.24 µs. Detailed horizontal timing is dependent on the specific video interface used, as discussed in Chapter 4.

Progressive Digital Component Video

SMPTE 296M specifies the representation for 720p digital R′G′B′ or YCbCr signals. Active resolutions defined within SMPTE 296M, their 1 × Y and R′G′B′ sample rates (Fs), and frame rates, are:

1280 × 720p	74.176 MHz	23.976 Hz
1280 × 720p	74.250 MHz	24.000 Hz
1280 × 720p	74.250 MHz	25.000 Hz
1280 × 720p	74.176 MHz	29.970 Hz
1280 × 720p	74.250 MHz	30.000 Hz

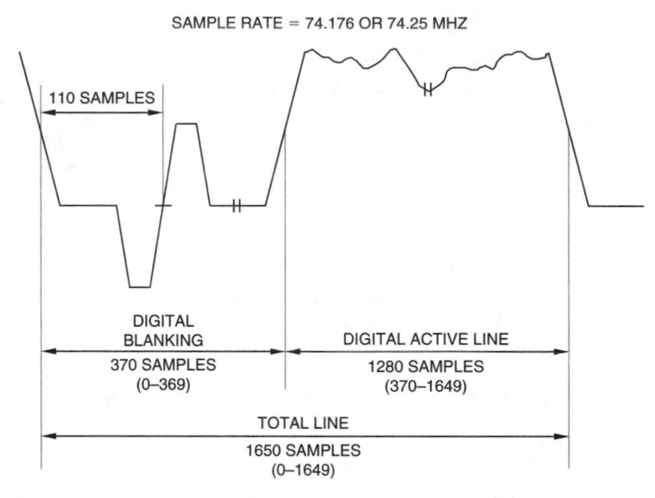

FIGURE 3.27 720p Analog-Digital Relationship (16:9 Aspect Ratio, 59.94 Hz Frame Rate, 74.176 MHz Sample Clock and 60 Hz Frame Rate, 74.25 MHz Sample Clock).

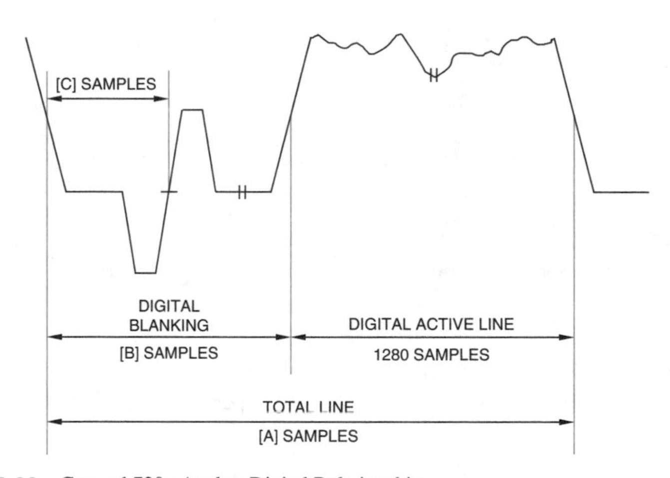

FIGURE 3.28 General 720p Analog-Digital Relationship.

1280 × 720p	74.250 MHz	50.000 Hz
1280 × 720p	74.176 MHz	59.940 Hz
1280 × 720p	74.250 MHz	60.000 Hz

Note that square pixels and a 16:9 aspect ratio are used. Example relationships between the analog and digital signals are shown in Figures 3.27 and 3.28, and Table 3.1. The H (horizontal blanking), V (vertical blanking), and F (field) signals are as defined in Figure 3.29.

TABLE 3.1 Various 720p Analog-Digital Parameters for Figure 3.28

Active Horizontal Samples	Frame Rate (Hz)	1× Y Sample Rate (MHz)	Total Horizontal Samples (A)	Horizontal Blanking Samples (B)	C Samples
1280	24/1.001	74.25/1.001	4125	2845	2585
	24	74.25	4125	2845	2585
	25[1]	48	1536	256	21
	25[1]	49.5	1584	304	25
	25	74.25	3960	2680	2420
	30/1.001	74.25/1.001	3300	2020	1760
	30	74.25	3300	2020	1760
	50	74.25	1980	700	440
	60/1.001	74.25/1.001	1650	370	110
	60	74.25	1650	370	110

Note: 1. Useful for CRT-based 50 Hz HDTVs based on a 31.250 kHz horizontal frequency. Sync pulses are −300 mV bi-level, rather than ±300 mV tri-level. 720p content scaled vertically to 1152i active scan lines; 1250i total scan lines instead of 750p.

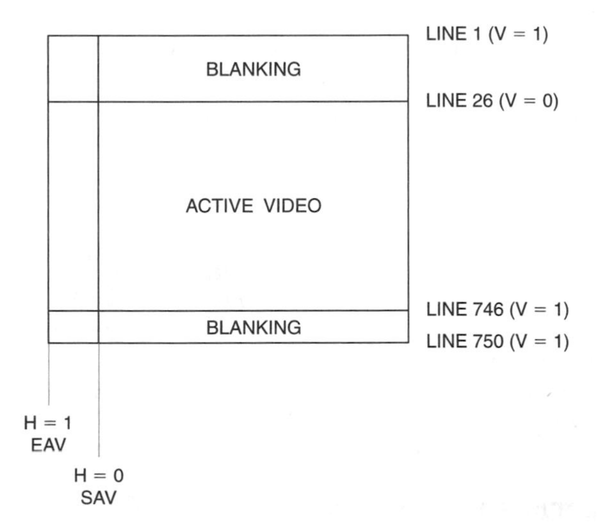

LINE NUMBER	F	V
1–25	0	1
26–745	0	0
746–750	0	1

FIGURE 3.29 720p Digital Vertical Timing (720 Active Lines). V changes state at the EAV sequence at the beginning of the digital line. Note that the digital line number changes state prior to the start of horizontal sync, as shown in Figures 3.27 and 3.28.

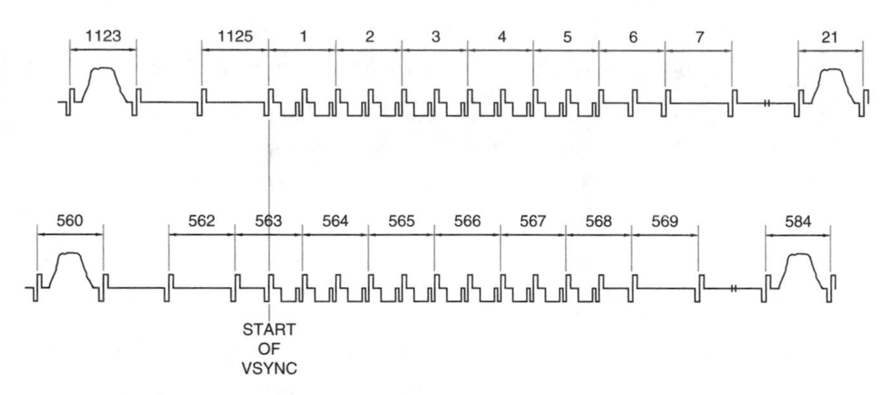

FIGURE 3.30 1080i Vertical Interval Timing.

1080i AND 1080p SYSTEMS

Interlaced Analog Component Video

Analog component signals are comprised of three signals, analog R′G′B′ or YPbPr. Referred to as 1080i (since there are typically 1080 active scan lines per frame and they are interlaced), the frame rate is usually 25 or 29.97 Hz (30/1.001) to simplify the generation of (B, D, G, H, I) PAL or (M) NTSC video. The analog interface uses 1125 lines per frame, with active video present on lines 21–560 and 584–1123, as shown in Figure 3.30.

MPEG-2 and MPEG-4 systems use 1088 lines, rather than 1080, in order to have a multiple of 32 scan lines per frame. In this case, an additional 4 lines per field after the active video are used.

For the 25 Hz frame rate, each scan line time is about 35.56 μs. For the 29.97 Hz frame rate, each scan line time is about 29.66 μs. Detailed horizontal timing is dependent on the specific video interface used, as discussed in Chapter 4.

Insider Info

*The **1152i** active (1250 total) line format is not a broadcast transmission format. However, it is being used as an analog interconnection standard from high-definition (HD) set-top boxes and DVD players to 50-Hz CRT-based HDTVs. This enables 50-Hz HDTVs to use a fixed 31.25 kHz horizontal frequency, reducing their cost. Other HDTV display technologies, such as DLP, LCD, and plasma, are capable of handling the native timing of 720p50 (750p50 with VBI) and 1080i25 (1125i25 with VBI) analog signals.*

Progressive Analog Component Video

Analog component signals are comprised of three signals, analog R′G′B′ or YPbPr. Referred to as 1080p (since there are typically 1080 active scan lines

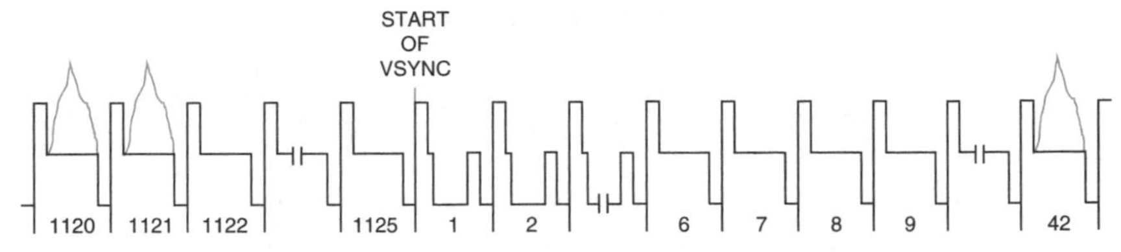

FIGURE 3.31 1080p Vertical Interval Timing.

per frame and they are progressive), the frame rate is usually 50 or 59.94 Hz (60/1.001) to simplify the generation of (B, D, G, H, I) PAL or (M) NTSC video. The analog interface uses 1125 lines per frame, with active video present on lines 42–1121, as shown in Figure 3.31.

MPEG-2 and MPEG-4 systems use 1088 lines, rather than 1080, in order to have a multiple of 16 scan lines per frame. In this case, an additional 8 lines per frame after the active video are used.

For the 50-Hz frame rate, each scan line time is about 17.78 μs. For the 59.94-Hz frame rate, each scan line time is about 14.83 μs. Detailed horizontal timing is dependent on the specific video interface used, as discussed in Chapter 4.

Interlaced Digital Component Video

ITU-R BT.709 and SMPTE 274M specify the digital component format for the 1080i digital R′G′B′ or YCbCr signal. Active resolutions defined within BT.709 and SMPTE 274M, their 1× Y and R′G′B′ sample rates (Fs), and frame rates, are:

1920 × 1080i	74.250 MHz	25.00 Hz
1920 × 1080i	74.176 MHz	29.97 Hz
1920 × 1080i	74.250 MHz	30.00 Hz

Note that square pixels and a 16:9 aspect ratio are used. Other common active resolutions, their 1× Y and R′G′B′ sample rates (Fs), and frame rates, are:

1280 × 1080i	49.500 MHz	25.00 Hz
1280 × 1080i	49.451 MHz	29.97 Hz
1280 × 1080i	49.500 MHz	30.00 Hz
1440 × 1080i	55.688 MHz	25.00 Hz
1440 × 1080i	55.632 MHz	29.97 Hz
1440 × 1080i	55.688 MHz	30.00 Hz

Example relationships between the analog and digital signals are shown in Figures 3.32 and 3.33, and Table 3.2. The H (horizontal blanking) and V (vertical blanking) signals are as defined in Figure 3.34.

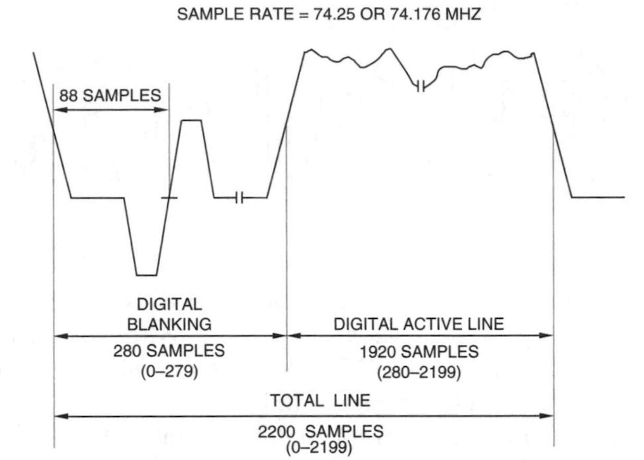

FIGURE 3.32 1080i Analog-Digital Relationship (16:9 Aspect Ratio, 29.97 Hz Frame Rate, 74.176 MHz Sample Clock and 30 Hz Frame Rate, 74.25 MHz Sample Clock).

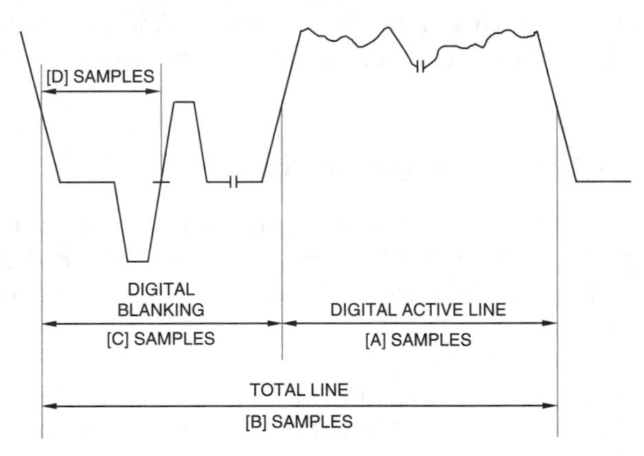

FIGURE 3.33 General 1080i Analog-Digital Relationship.

Progressive Digital Component Video

ITU-R BT.709 and SMPTE 274M specify the digital component format for the 1080p digital R′G′B′ or YCbCr signal. Active resolutions defined within BT.709 and SMPTE 274M, their 1× Y and R′G′B′ sample rates (Fs), and frame rates, are:

1920 × 1080p	74.176 MHz	23.976 Hz
1920 × 1080p	74.250 MHz	24.000 Hz
1920 × 1080p	74.250 MHz	25.000 Hz
1920 × 1080p	74.176 MHz	29.970 Hz
1920 × 1080p	74.250 MHz	30.000 Hz
1920 × 1080p	148.50 MHz	50.000 Hz

TABLE 3.2 Various 1080i analog-digital parameters for Figure 3.33

Active Horizontal Samples (A)	Frame Rate (Hz)	1× Y Sample Rate (MHz)	Total Horizontal Samples (B)	Horizontal Blanking Samples (C)	D Samples
1920	25[1]	72	2304	384	32
	25[1]	74.25	2376	456	38
	25	74.25	2640	720	528
	30/1.001	74.25/1.001	2200	280	88
	30	74.25	2200	280	88
1440	25[1]	54	1728	288	24
	25	55.6875	1980	540	396
	30/1.001	55.6875/ 1.001	1650	210	66
	30	55.6875	1650	210	66
1280	25[1]	48	1536	256	21
	25	49.5	1760	480	352
	30/1.001	49.5/1.001	1466.7	186.7	58.7
	30	49.5	1466.7	186.7	58.7

Notes: 1. Useful for CRT-based 50 Hz HDTVs based on a 31.250 kHz horizontal frequency. Sync pulses are −300 mV bi-level, rather than ±300 mV tri-level. 1080i content letterboxed in 1152i active scan lines; 1250i total scan lines instead of 1125i.

1920 × 1080p	148.35 MHz	59.940 Hz
1920 × 1080p	148.50 MHz	60.000 Hz

Note that square pixels and a 16:9 aspect ratio are used. Other common active resolutions, their 1× Y and R′G′B′ sample rates (Fs), and frame rates, are:

1280 × 1080p	49.451 MHz	23.976 Hz
1280 × 1080p	49.500 MHz	24.000 Hz
1280 × 1080p	49.500 MHz	25.000 Hz
1280 × 1080p	49.451 MHz	29.970 Hz
1280 × 1080p	49.500 MHz	30.000 Hz
1280 × 1080p	99.000 MHz	50.000 Hz
1280 × 1080p	98.901 MHz	59.940 Hz
1280 × 1080p	99.000 MHz	60.000 Hz
1440 × 1080p	55.632 MHz	23.976 Hz
1440 × 1080p	55.688 MHz	24.000 Hz
1440 × 1080p	55.688 MHz	25.000 Hz
1440 × 1080p	55.632 MHz	29.970 Hz
1440 × 1080p	55.688 MHz	30.000 Hz
1440 × 1080p	111.38 MHz	50.000 Hz

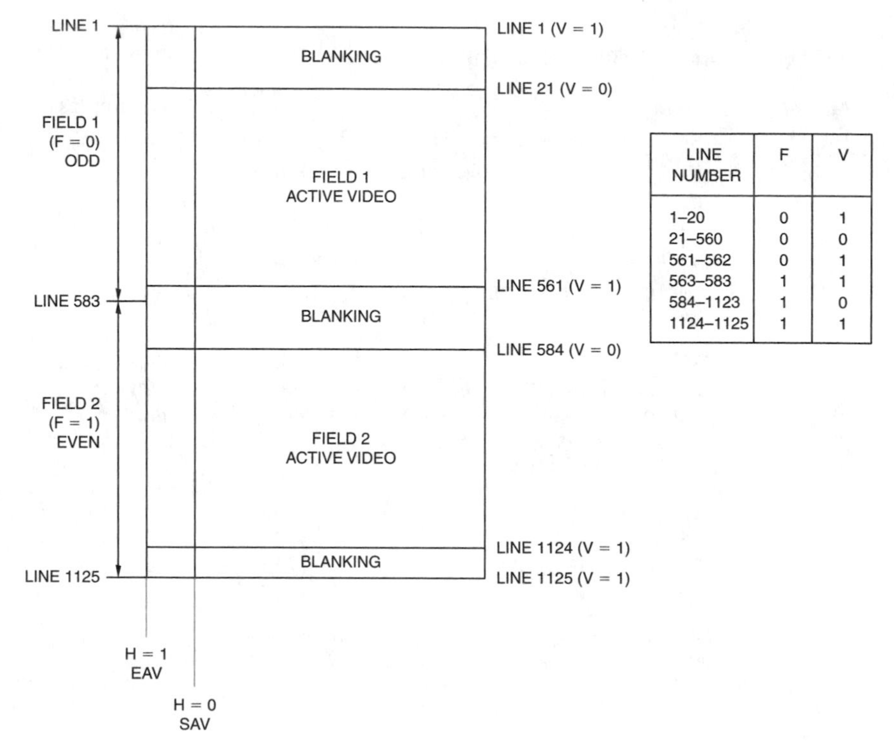

LINE NUMBER	F	V
1–20	0	1
21–560	0	0
561–562	0	1
563–583	1	1
584–1123	1	0
1124–1125	1	1

FIGURE 3.34 1080i Digital Vertical Timing (1080 Active Lines). F and V change state at the EAV sequence at the beginning of the digital line. Note that the digital line number changes state prior to the start of horizontal sync, as shown in Figures 3.32 and 3.33.

1440 × 1080p	111.26 MHz	59.940 Hz
1440 × 1080p	111.38 MHz	60.000 Hz

Example relationships between the analog and digital signals are shown in Figures 3.35 and 3.36, and Table 3.3. The H (horizontal blanking), V (vertical blanking), and F (field) signals are as defined in Figure 3.37.

Insider Info

Some consumer displays, such as those based on LCD and plasma technologies, have adapted other resolutions as their native resolution. Common active resolutions and their names are:

640 × 400	VGA
640 × 480	VGA
854 × 480	WVGA
800 × 600	SVGA

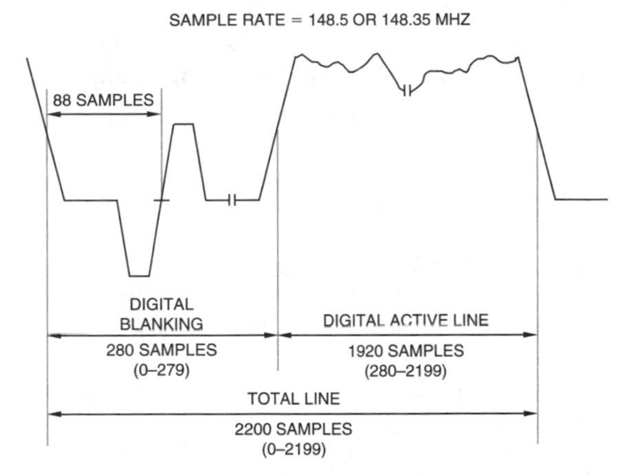

FIGURE 3.35 1080p Analog-Digital Relationship (16:9 Aspect Ratio, 59.94 Hz Frame Rate, 148.35 MHz Sample Clock and 60 Hz Frame Rate, 148.5 MHz Sample Clock).

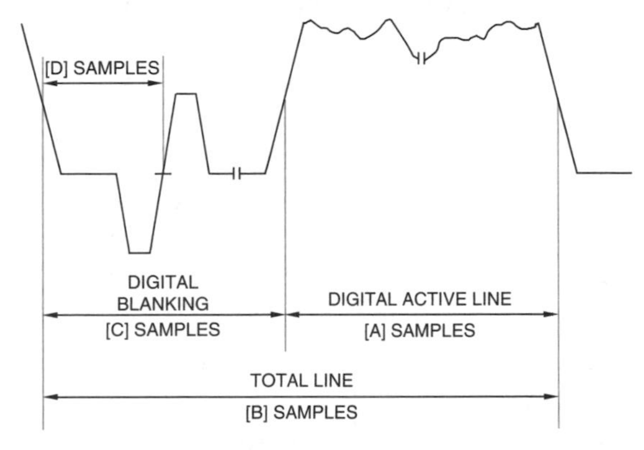

FIGURE 3.36 General 1080p Analog-Digital Relationship.

1024 × 768	XGA
1280 × 768	WXGA
1366 × 768	WXGA
1024 × 1024	XGA
1280 × 1024	SXGA
1600 × 1024	WSXGA
1600 × 1200	UXGA
1920 × 1200	WUXGA

These resolutions, and their timings, are defined for computer monitors by the Video Electronics Standards Association (VESA). Displays based on one of these native resolutions are usually capable of accepting many input resolutions, scaling the source to match the display resolution.

TABLE 3.3 Various 1080p Analog-Digital Parameters for Figure 3.36

Active Horizontal Samples (A)	Frame Rate (Hz)	1× Y Sample Rate (MHz)	Total Horizontal Samples (B)	Horizontal Blanking Samples (C)	D Samples
1920	24/1.001	74.25/1.001	2750	830	638
	24	74.25	2750	830	638
	25	74.25	2640	720	528
	30/1.001	74.25/1.001	2200	280	88
	30	74.25	2200	280	88
	50	148.5	2640	720	528
	60/1.001	148.5/1.001	2200	280	88
	60	148.5	2200	280	88
1440	24/1.001	55.6875/1.001	2062.5	622.5	478.5
	24	55.6875	2062.5	622.5	478.5
	25	55.6875	1980	540	396
	30/1.001	55.6875/1.001	1650	210	66
	30	55.6875	1650	210	66
	50	111.375	1980	540	396
	60/1.001	111.375/1.001	1650	210	66
	60	111.375	1650	210	66
1280	24/1.001	49.5/1.001	1833.3	553.3	425.3
	24	49.5	1833.3	553.3	425.3
	25	49.5	1760	480	352
	30/1.001	49.5/1.001	1466.7	186.7	58.7
	30	49.5	1466.7	186.7	58.7
	50	99	1760	480	352
	60/1.001	99/1.001	1466.7	186.7	58.7
	60	99	1466.7	186.7	58.7

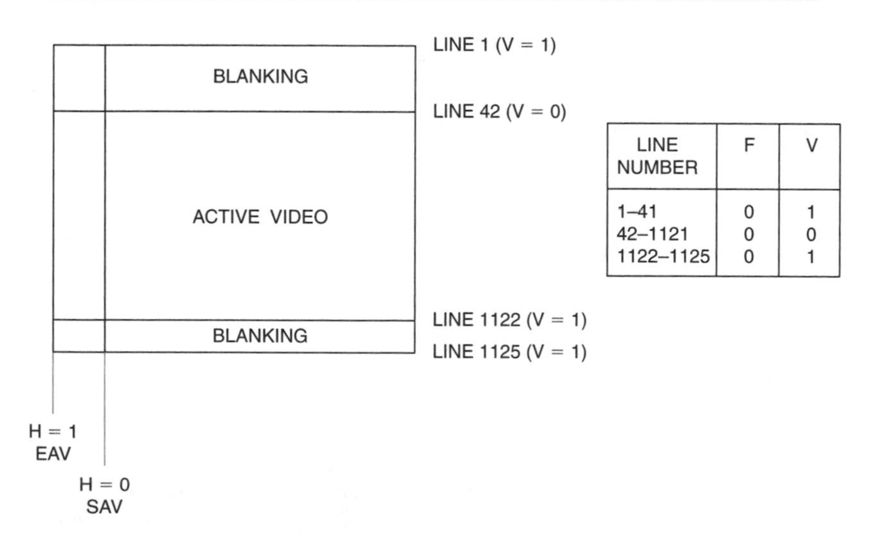

FIGURE 3.37 1080p Digital Vertical Timing (1080 Active Lines). V changes state at the EAV sequence at the beginning of the digital line. Note that the digital line number changes state prior to the start of horizontal sync, as shown in Figures 3.35 and 3.36.

INSTANT SUMMARY

Video signals come in many "flavors," with a wide variety of options. These include:

- Digital component video, developed by the EBU to support simpler international program exchange
- 480i and 480p systems, which include interlaced analog composite video, interlaced analog component video, progressive analog component video, interlaced digital component video, and progressive digital component video at 480 active scan lines per frame
- 576i and 576p systems, which include analog composite video, interlaced analog component video, progressive analog component video, interlaced digital component video, and progressive digital component video at 576 active scan lines per frame
- 720p systems, which include progressive analog component video and progressive digital component video at 720 active scan lines per frame
- 1080i and 1080p systems, which include interlaced analog component video, progressive analog component video, interlaced digital component video, and progressive digital component video at 1080 active scan lines per frame

Chapter 4

Video Interfaces

In an Instant

Analog video interface

- Definitions
- S-Video Interface
- SCART Interface
- SDTV RGB Interface
- HDTV RGB Interface
- SDTV YPbPr Interface
- HDTV YPbPr Interface
- D-Connector Interface
- Other Pro-Video analog Interfaces
- VGA Interface

Digital video interfaces

- Digital video interface definitions
- Pro-Video Component Interfaces
- Pro-Video Composite Interfaces
- Pro-Video Transport Interfaces
- IC Component Interfaces
- Consumer Component Interfaces
- Consumer Transport Interfaces

ANALOG VIDEO INTERFACES

Definitions

For years, the primary video signal used by the consumer market has been composite NTSC or PAL video. Attempts have been made to support S-video, but, until recently, it has been largely limited to S-VHS VCRs and high-end televisions. However, with the introduction of DVD players, digital set-top boxes, and DTV, there has been renewed interest in providing high-quality video to the consumer market. These devices not only support very high-quality composite and S-video signals, but many of them also allow the option of using analog R´G´B´ or YPbPr video. In this section, we will cover analog video interfaces. First we'll define some terms used in this section.

VBI, or *vertical blanking interval*, refers to the period when the video signal is at the blanking level so as not to display the electron beam when it sweeps back from the bottom to the top side of the CRT screen.

HDMI (High-Definition Multimedia Interface) is a single-cable digital audio/video interface for consumer equipment. It is designed to replace DVI in a backwards-compatible fashion.

DVI (Digital Visual Interface) is a digital video interface to a display, designed to replace the analog YPbPr or R′G′B′ interface.

Sync is a fundamental piece of information for displaying any type of video. Essentially, the sync signal tells the display where to put the picture. Horizontal sync (HSYNC) tells the display where to put the picture in the left-to-right dimension. The vertical sync (VSYNC) tells the display where to put the picture from top to bottom.

IRE refers to an arbitrary unit used to describe the amplitude characteristics of a video signal. White is defined to be 100 IRE and the blanking level is defined to be 0 IRE.

Technology Trade-offs

Using analog R′G′B′ or YPbPr video eliminates NTSC/PAL encoding and decoding artifacts. As a result, the picture is sharper and has less noise. More color bandwidth is also available, increasing the horizontal detail.

S-Video Interface

The RCA phono connector (consumer market) or BNC connector (pro-video market) transfers a composite NTSC or PAL video signal, made by adding the intensity (Y) and color (C) video signals together. The television then has to separate these Y and C video signals in order to display the picture. The problem is that the Y/C separation process is never perfect.

Many video components now support a 4-pin "S1" S-video connector, illustrated in Figure 4.1 (the female connector viewpoint). This connector keeps the intensity (Y) and color (C) video signals separate, eliminating the Y/C separation process in the TV. As a result, the picture is sharper and has less noise.

NTSC and PAL VBI (vertical blanking interval) data may be present on the 480i or 576i Y video signal.

The "S2" version adds a +5 V DC offset to the C signal when a widescreen (16:9) anamorphic program (horizontally squeezed by 25%) is present. A 16:9 TV detects the DC offset and horizontally expands the 4:3 image to fill the screen, restoring the correct aspect ratio of the program. The "S3" version also supports using a +2.3 V offset when a program is letterboxed.

The IEC 60933-5 standard specifies the S-video connector, including signal levels.

Insider Info

The PC market also uses an extended S-Video interface. This interface has 7 pins, as shown in Figure 4.1, and is backwards compatible with the 4-pin interface. The use of the three additional pins varies by manufacturer. They may be used to support an I2C interface (SDA bidirectional data pin and SCL clock pin), +12V power, a composite NTSC/PAL video signal (CVBS), or analog R′G′B′ or YPbPr video signals.

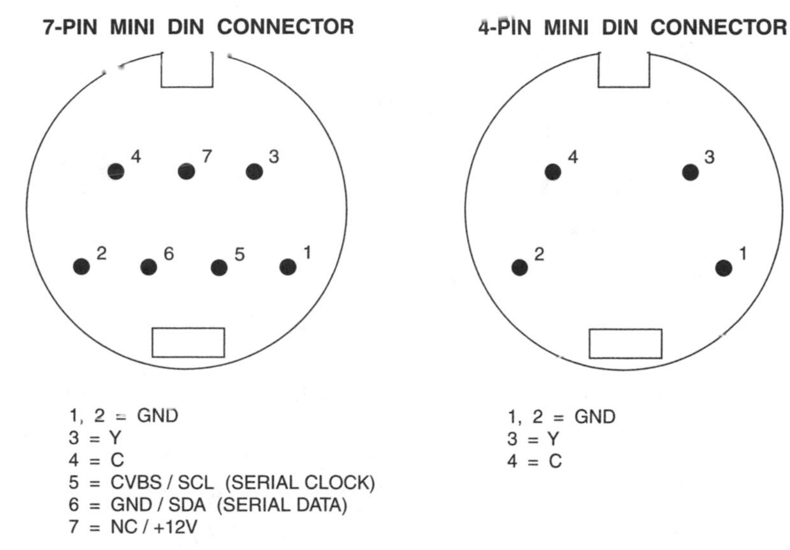

FIGURE 4.1 S-Video Connector and Signal Names.

SCART Interface

Most consumer video components in Europe support one or two 21-pin SCART connectors (also known as Peritel, Peritelevision, and Euroconnector). This connection allows analog R′G′B′ video or S-video, composite video, and analog stereo audio to be transmitted between equipment using a single cable. The composite video signal must always be present, as it provides the basic video timing for the analog R′G′B′ video signals. Note that the 700 mV R′G′B′ signals do not have a blanking pedestal or sync information, as illustrated in Figure 4.4.

PAL VBI (vertical blanking interval) data may be present on the 576i composite video signal.

There are now several types of SCART pinouts, depending on the specific functions implemented, as shown in Table 4.1. Pinout details are shown in Figure 4.2.

The CENELEC EN 50049–1 and IEC 60933 standards specify the basic SCART connector, including signal levels.

SDTV RGB Interface

Some SDTV consumer video equipment supports an analog R′G′B′ video interface. NTSC and PAL VBI data may be present on 480i or 576i R′G′B′ video signals. Three separate RCA phono connectors (consumer market) or BNC connectors (pro-video and PC market) are used.

The horizontal and vertical video timing are dependent on the video standard, as discussed in Chapter 3. For sources, the video signal at the connector should

TABLE 4.1 SCART Connector Signals

Pin	Function	Signal Level	Impedance
1	right audio out	0.5V rms	< 1K ohm
2	right audio in	0.5V rms	> 10K ohm
3	left/mono audio out	0.5V rms	< 1K ohm
4	ground - for pins 1, 2, 3, 6		
5	ground - for pin 7		
6	left/mono audio in	0.5V rms	> 10K ohm
7	blue (or C) video in/out	0.7V (or 0.3V burst)	75 ohms
8	status and aspect ratio in/out	9.5V–12V = 4:3 source 4.5V–7V = 16:9 source 0V–2V = inactive source	> 10K ohm
9	ground - for pin 11		
10	data 2		
11	green video in/out	0.7V	75 ohms
12	data 1		
13	ground - for pin 15		
14	ground - for pin 16		
15	red (or C) video in/out	0.7V (or 0.3V burst)	75 ohms
16	RGB control in/out	1–3V = RGB, 0–0.4V = composite	75 ohms
17	ground - for pin 19		
18	ground - for pin 20		
19	composite (or Y) video out	1V	75 ohms
20	composite (or Y) video in	1V	75 ohms
21	ground - for pins 8, 10, 12, shield		

Note: Often, the SCART 1 connector supports composite video and RGB, the SCART 2 connector supports composite video and S-Video, and the SCART 3 connector supports only composite video. SCART connections may also be used to add external decoders or descramblers to the video path, the video signal goes out and comes back in.

The RGB control signal controls the TV switch between the composite and RGB inputs, enabling the overlaying of text onto the video, even the internal TV program. This enables an external teletext or closed captioning decoder to add information over the current program. If pin 16 is held high, signifying RGB signals are present, the sync is still carried on the Composite Video pin. Some devices (such as DVD players) may provide RGB on a SCART and hold pin 16 permanently high.

When a source becomes active, it sets a 12V level on pin 8. This causes the TV to automatically switch to that SCART input. When the source stops, the signal returns to 0V and TV viewing is resumed. If an anamorphic 16:9 program is present, the source raises the signal on pin 8 to only 6V. This causes the TV to switch to that SCART input and at the same time enable the video processing for anamorphic 16:9 programs.

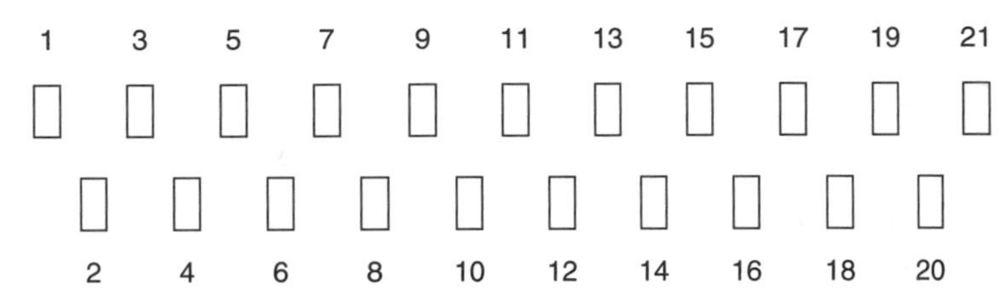

FIGURE 4.2 SCART Connector.

have a source impedance of 75Ω ±5%. For receivers, vidco inputs should be AC-coupled and have a 75Ω ±5% input impedance. The three signals must be coincident with respect to each other within ±5 ns.

Sync information may be present on just the green channel, all three channels, as a separate composite sync signal, or as separate horizontal and vertical sync signals. A gamma of 1/0.45 is used.

7.5 IRE Blanking Pedestal

As shown in Figure 4.3, the nominal active video amplitude is 714 mV, including a 7.5 ±2 IRE blanking pedestal. A 286 ±6 mV composite sync signal may be present on just the green channel (consumer market), or all three channels (pro-video market). DC offsets up to ±1 V may be present.

Table 4.2 shows SDTV 10-bit R′G′B′ values.

0 IRE Blanking Pedestal

As shown in Figure 4.4, the nominal active video amplitude is 700 mV, with no blanking pedestal. A 300±6 mV composite sync signal may be present on just the green channel (consumer market), or all three channels (pro-video market). DC offsets up to ±1 V may be present.

HDTV RGB Interface

Some HDTV consumer video equipment supports an analog R′G′B′ video interface. Three separate RCA phono connectors (consumer market) or BNC connectors (pro-video and PC market) are used.

The horizontal and vertical video timing are dependent on the video standard, as discussed in Chapter 3. For sources, the video signal at the connector should have a source impedance of 75 Ω±5%. For receivers, video inputs should be AC-coupled and have a 75 Ω±5% input impedance. The three signals must be coincident with respect to each other within ±5 ns.

Sync information may be present on just the green channel, all three channels, as a separate composite sync signal, or as separate horizontal and vertical sync signals. A gamma of 1/0.45 is used.

As shown in Figure 4.5, the nominal active video amplitude is 700 mV, and has no blanking pedestal. A ±300 ±6-mV tri-level composite sync signal may

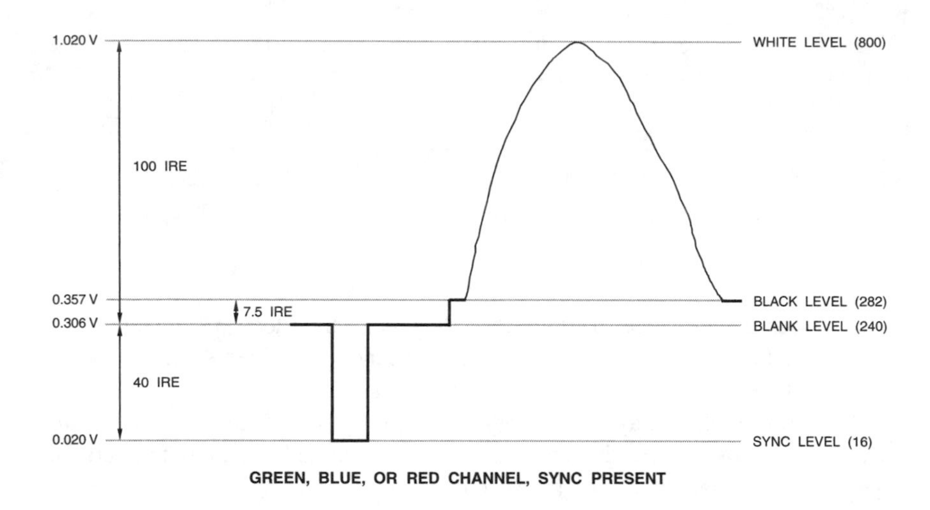

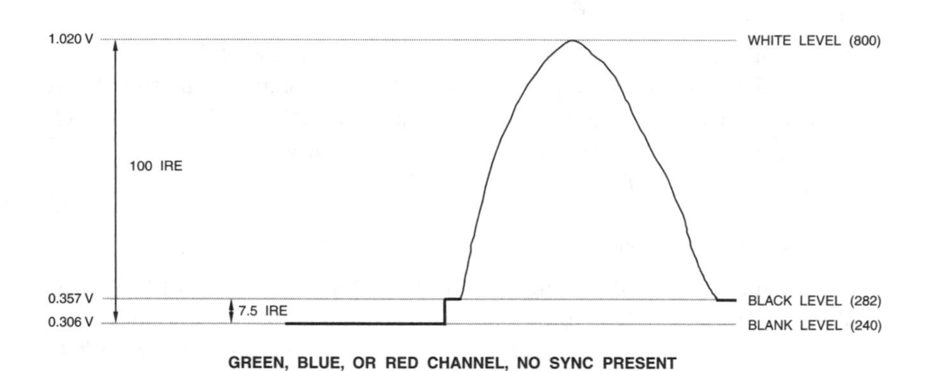

FIGURE 4.3 SDTV Analog RGB Levels. 7.5 IRE blanking level.

be present on just the green channel (consumer market), or all three channels (pro video market). DC offsets up to ±1 V may be present.

Table 4.3 shows HDTV 10-bit R′G′B′ values.

Insider Info

Due to the limited availability of copy protection technology for high-definition analog interfaces, some standards and DRM implementations only allow a constrained image to be output. A constrained image has an effective maximum resolution of 960 × 540p, although the total number of video samples and the video timing remain unchanged (for example, 1280 × 720p or 1920 × 1080i). In these situations, the full resolution image is still available via an approved secure digital video output, such as HDMI.

TABLE 4.2 SDTV 10-Bit R′G′B′ Values

Video Level	7.5 IRE Blanking Pedestal	0 IRE Blanking Pedestal
White	800	800
Black	282	252
Blank	240	252
Sync	16	16

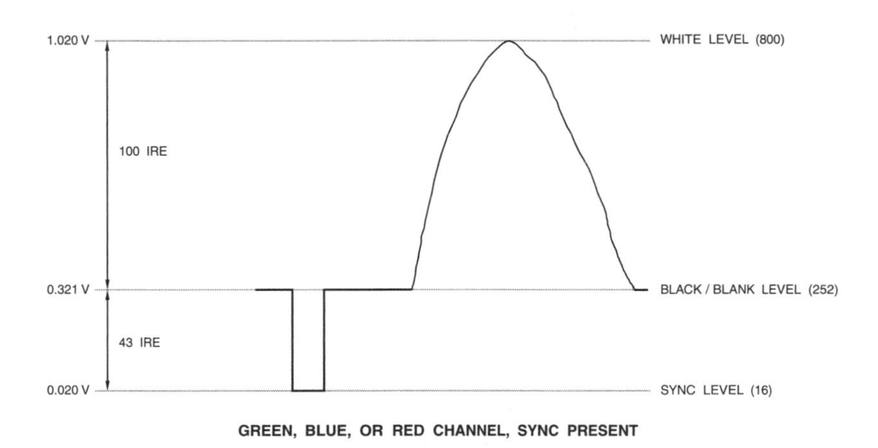

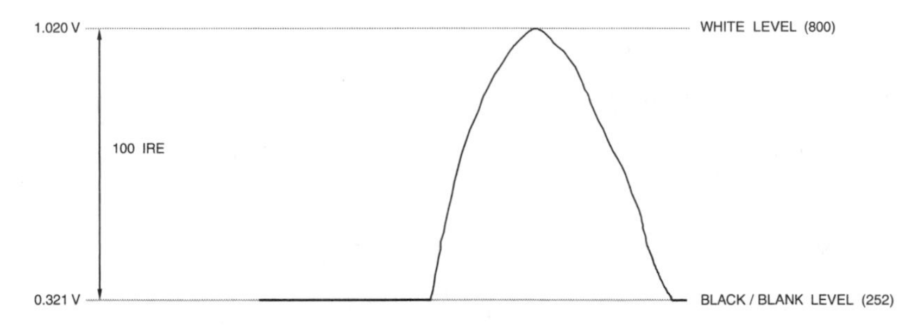

FIGURE 4.4 SDTV Analog RGB Levels. 0 IRE blanking level.

SDTV YPbPr Interface

Some SDTV consumer video equipment supports an analog YPbPr video interface. NTSC and PAL VBI (vertical blanking interval) data may be present on 480i or 576i Y video signals. Three separate RCA phono connectors (consumer market) or BNC connectors (pro-video market) are used.

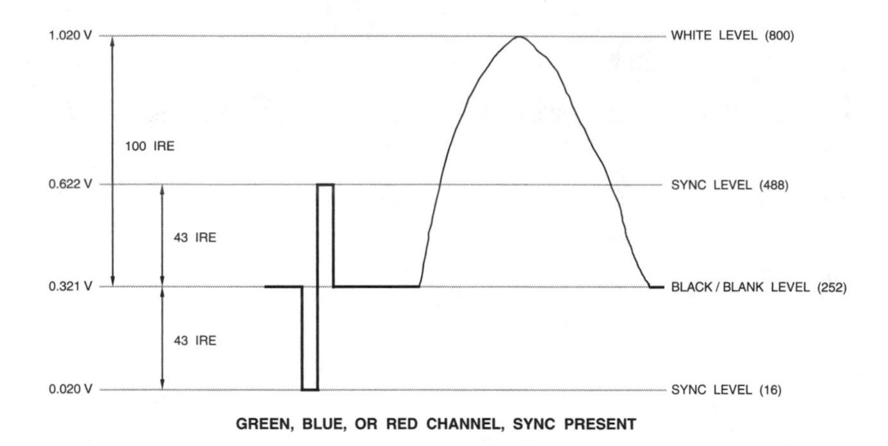

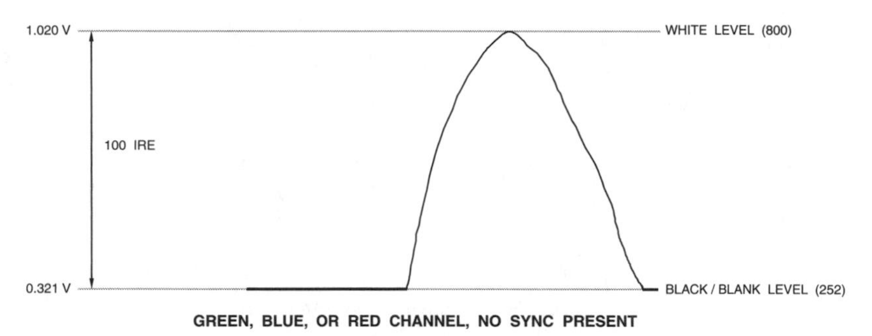

FIGURE 4.5 HDTV Analog RGB Levels. 0 IRE blanking level.

TABLE 4.3 HDTV 10-Bit R′G′B′ Values

Video Level	0 IRE Blanking Pedestal
White	800
Sync high	488
Black	252
Blank	252
Sync low	16

The horizontal and vertical video timing are dependent on the video standard, as discussed in Chapter 3. For sources, the video signal at the connector should have a source impedance of $75\,\Omega \pm 5\%$. For receivers, video inputs should be AC-coupled and have a $75\,\Omega \pm 5\%$ input impedance. The three signals must be coincident with respect to each other within ± 5 ns.

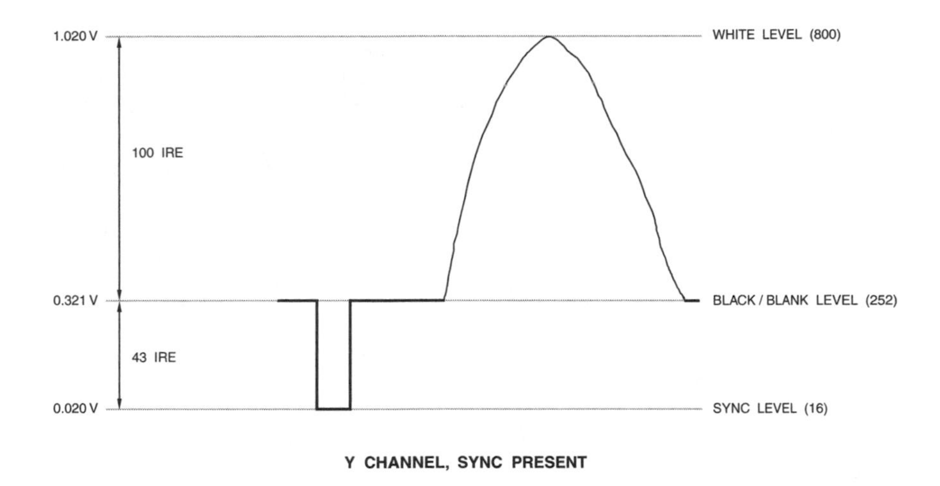

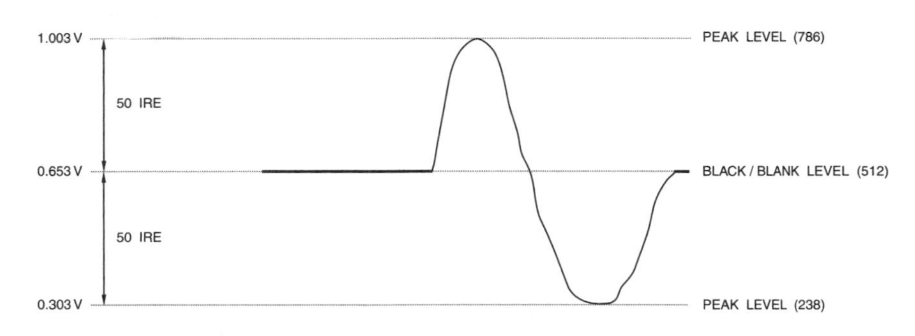

FIGURE 4.6 EIA-770.2 SDTV Analog YPbPr Levels. Sync on Y.

For consumer products, composite sync is present on only the Y channel. For pro-video applications, composite sync is present on all three channels. A gamma of 1/0.45 is specified.

As shown in Figures 4.6 and 4.7, the Y signal consists of 700 mV of active video (with no blanking pedestal). Pb and Pr have a peak-to-peak amplitude of 700 mV. A 300 ±6 mV composite sync signal is present on just the Y channel (consumer market), or all three channels (pro-video market). DC offsets up to ±1 V may be present. The 100% and 75% YPbPr color bar values are shown in Tables 4.4 and 4.5. Table 4.6 shows SDTV 10-bit YPbPr values.

HDTV YPbPr Interface

Most HDTV consumer video equipment supports an analog YPbPr video interface. Three separate RCA phono connectors (consumer market) or BNC connectors (pro-video market) are used.

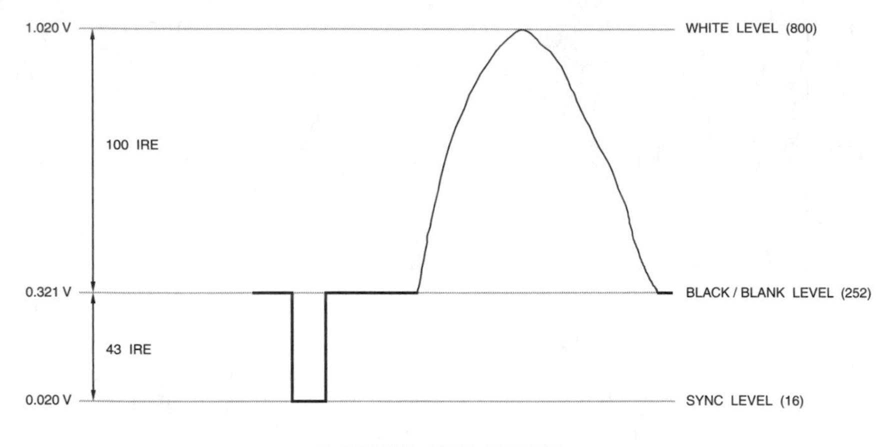

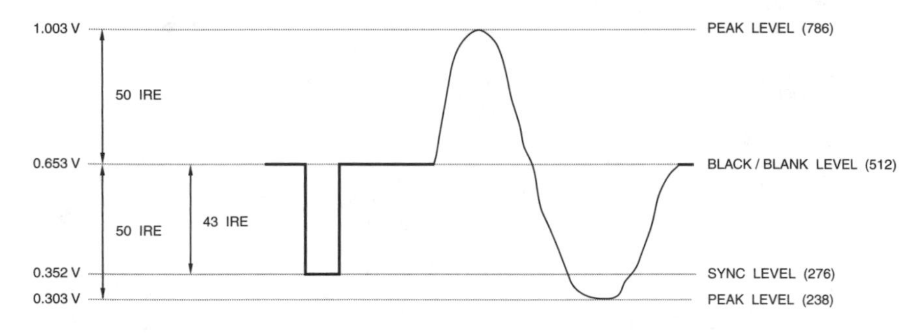

FIGURE 4.7 SDTV Analog YPbPr Levels. Sync on YPbPr.

The horizontal and vertical video timing is dependent on the video standard, as discussed in Chapter 3. For sources, the video signal at the connector should have a source impedance of $75\,\Omega \pm 5\%$. For receivers, video inputs should be AC-coupled and have a $75\,\Omega \pm 5\%$ input impedance. The three signals must be coincident with respect to each other within ± 5 ns.

For consumer products, composite sync is present on only the Y channel. For pro-video applications, composite sync is present on all three channels. A gamma of 1/0.45 is specified.

As shown in Figures 4.8 and 4.9, the Y signal consists of 700 mV of active video (with no blanking pedestal). Pb and Pr have a peak-to-peak amplitude of 700 mV. A ±300 ±6 mV composite sync signal is present on just the Y channel (consumer market), or all three channels (pro-video market). DC offsets up to ±1 V may be present. The 100% and 75% YPbPr color bar values are shown in Tables 4.7 and 4.8.

Table 4.9 shows HDTV 10-bit YPbPr values.

TABLE 4.4 EIA-770.2 SDTV YPbPr and YCbCr 100% Color Bars. YPbPr values relative to the blanking level

		White	Yellow	Cyan	Green	Magenta	Red	Blue	Black
Y	IRE	100	88.6	70.1	58.7	41.3	29.9	11.4	0
	mV	700	620	491	411	289	209	80	0
Pb	IRE	0	−50	16.8	−33.1	33.1	−16.8	50	0
	mV	0	−350	118	−232	232	−118	350	0
Pr	IRE	0	8.1	−50	−41.8	41.8	50	−8.1	0
	mV	0	57	−350	−293	293	350	−57	0
Y	64 to 940	940	840	678	578	426	326	164	64
Cb	64 to 960	512	64	663	215	809	361	960	512
Cr	64 to 960	512	585	64	137	887	960	439	512

TABLE 4.5 EIA-770.2 SDTV YPbPr and YCbCr 75% Color Bars. YPbPr values relative to the blanking level

		White	Yellow	Cyan	Green	Magenta	Red	Blue	Black
Y	IRE	75	66.5	52.6	44	31	22.4	8.6	0
	mV	525	465	368	308	217	157	60	0
Pb	IRE	0	−37.5	12.6	−24.9	24.9	−12.6	37.5	0
	mV	0	−262	88	−174	174	–88	262	0
Pr	IRE	0	6.1	−37.5	−31.4	31.4	37.5	−6.1	0
	mV	0	43	−262	−220	220	262	−43	0
Y	64 to 940	721	646	525	450	335	260	139	64
Cb	64 to 960	512	176	625	289	735	399	848	512
Cr	64 to 960	512	567	176	231	793	848	457	512

TABLE 4.6 SDTV 10-Bit YPbPr Values

Video Level	Y	PbPr
White	800	512
Black	252	512
Blank	252	512
Sync	16	276

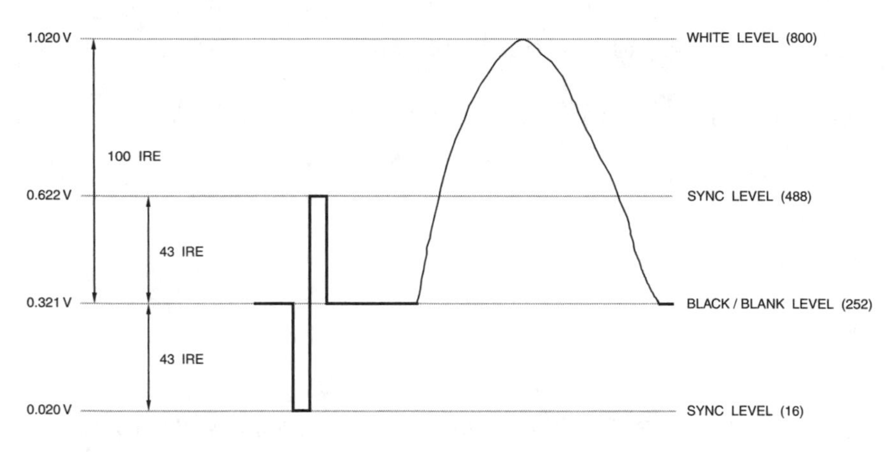

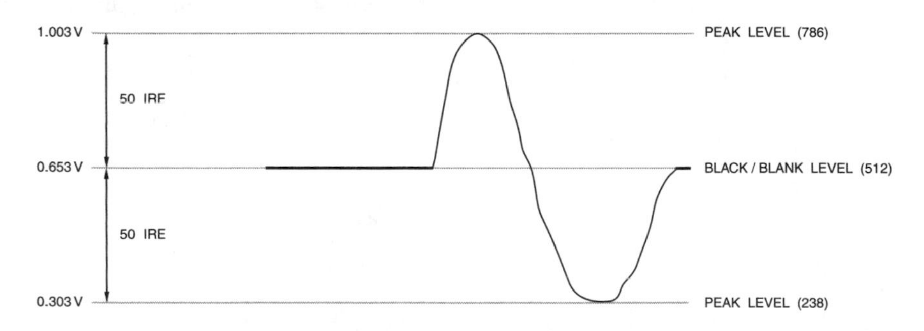

FIGURE 4.8 EIA-770.3 HDTV Analog YPbPr Levels. Sync on Y.

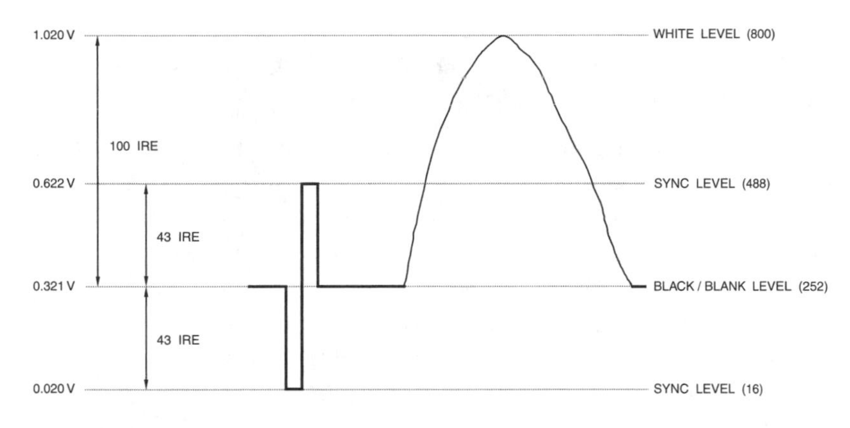

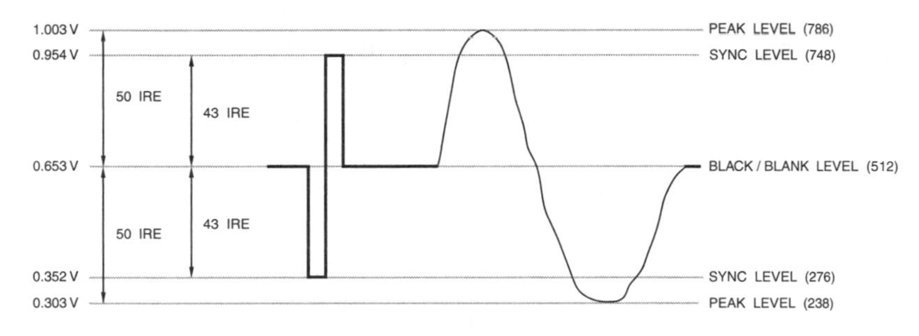

FIGURE 4.9 SMPTE 274M and 296M HDTV Analog YPbPr Levels. Sync on YPbPr.

Insider Info

Due to the limited availability of copy protection technology for high-definition analog interfaces, some standards and DRM implementations only allow a constrained image to be output. A constrained image has an effective maximum resolution of 960 × 540p, although the total number of video samples and the video timing remain unchanged (for example, 1280 × 720p60 or 1920 × 1080i30). In these situations, the full resolution image is still available via an approved secure digital video output, such as HDMI.

D-Connector Interface

A 14-pin female D-connector (EIA-J CP–4120 standard, EIA-J RC–5237 connector) is optionally used on some high-end consumer equipment in Japan, Hong Kong, and Singapore. It is used to transfer EIA 770.2 or EIA 770.3 interlaced or progressive analog YPbPr video.

TABLE 4.7 EIA-770.3 HDTV YPbPr and YCbCr 100% Color Bars. YPbPr values relative to the blanking level

		White	Yellow	Cyan	Green	Magenta	Red	Blue	Black
Y	IRE	100	92.8	78.7	71.5	28.5	21.3	7.2	0
	mV	700	650	551	501	200	149	50	0
Pb	IRE	0	−50	11.4	−38.5	38.5	−11.4	50	0
	mV	0	−350	80	−270	270	−80	350	0
Pr	IRE	0	4.6	−50	−45.4	45.4	50	−4.6	0
	mV	0	32	−350	−318	318	350	−32	0
Y	64 to 940	940	877	753	690	314	251	127	64
Cb	64 to 960	512	64	614	167	857	410	960	512
Cr	64 to 960	512	553	64	106	918	960	471	512

TABLE 4.8 EIA-770.3 HDTV YPbPr and YCbCr 75% Color Bars. YPbPr values relative to the blanking level

		White	Yellow	Cyan	Green	Magenta	Red	Blue	Black
Y	IRE	75	69.6	59	53.7	21.3	16	5.4	0
	mV	525	487	413	376	149	112	38	0
Pb	IRE	0	−37.5	8.6	−28.9	28.9	−8.6	37.5	0
	mV	0	−263	60	−202	202	−60	263	0
Pr	IRE	0	3.5	−37.5	−34	34	37.5	−3.5	0
	mV	0	24	−263	−238	238	263	−24	0
Y	64 to 940	721	674	581	534	251	204	111	64
Cb	64 to 960	512	176	589	253	771	435	848	512
Cr	64 to 960	512	543	176	207	817	848	481	512

TABLE 4.9 HDTV 10-Bit YPbPr Values

Video Level	Y	PbPr
White	800	512
Sync high	488	748
Black	252	512
Blank	252	512
Sync low	16	276

TABLE 4.10 D-Connector Supported Video Formats

	480i	480p	720p	1080i	1080p
D1	×				
D2	×	×			
D3	×	×		×	
D4	×	×	×	×	
D5	×	×	×	×	×

There are five flavors of the D-connector, referred to as D1, D2, D3, D4, and D5, each used to indicate supported video formats, as shown in Table 4.10. Figure 4.10 illustrates the connector and Table 4.11 lists the pin names.

Three line signals (Line 1, Line 2, and Line 3) indicate the resolution and frame rate of the YPbPr source video, as indicated in Table 4.12.

Other Pro-Video Analog Interfaces

Tables 4.13 and 4.14 list some other common component analog video formats. The horizontal and vertical timing is the same as for 525-line (M) NTSC

TABLE 4.11 D-Connector Pin Descriptions

Pin	Function	Signal Level	Impedance
1	Y	0.700V + sync	75 ohms
2	ground - Y		
3	Pb	±0.350V	75 ohms
4	ground - Pb		
5	Pr	±0.350V	75 ohms
6	ground - Pr		
7	reserved 1		
8	line 1	0V, 2.2V, or 5V[1]	10K±3K ohm
9	line 2	0V, 2.2V, or 5V[1]	10K±3K ohm
10	reserved 2		
11	line 3	0V, 2.2V, or 5V[1]	10K±3K ohm
12	ground - detect plugged		
13	reserved 3		
14	detect plugged	0V = plugged in[2]	> 100K ohm

Notes: 1. 2.2V has range of 2.2V ±0.8V. 5V has a range of 5V ±1.5V.
2. Inside equipment, pin 12 is connected to ground and pin 14 is pulled to 5V through a resistor. Inside each D-Connector plug, pins 12 and 14 are shorted together.

and 625-line (B, D, G, H, I) PAL. The 100% and 75% color bar values are shown in Tables 4.15 through 4.18. The SMPTE, EBU N10, 625-line Betacam, and 625-line MII values are the same as for SDTV YPbPr.

VGA Interface

Table 4.19 and Figure 4.11 illustrate the 15-pin VGA connector used by computer equipment, and some consumer equipment, to transfer analog RGB signals. The analog RGB signals do not contain sync information and have no blanking pedestal, as shown in Figure 4.4.

DIGITAL VIDEO INTERFACES

Digital Video Interface Definitions

This section will cover interfaces for digital video. First we will define some terms used in the sections to follow.

TABLE 4.12 Voltage Levels of Line Signals for Various Video Formats for D-Connector.

Resolution		Frame Rate	Line 1 Scan Lines	Line 2 Frame Rate	Line 3 Aspect Ratio	Chromaticity and Reference White	Color Space Equations	Gamma Correction	Sync Amplitude on Y
1920 × 1080		**30i**	5V	0V	5V	EIA-770.3	EIA-770.3	EIA-770.3	±0.300V[3]
		25i[2]	5V	2.2V	5V				
		30p	5V	2.2V	5V				
		25p[2]	5V	2.2V	5V				
		24p[2]	5V	2.2V	5V				
		24sF[2]	5V	2.2V	5V				
1280 × 720		**60p**	2.2V	5V	5V				
		50p[2]	2.2V	2.2V	5V				
		30p	2.2V	2.2V	5V				
		25p[2]	2.2V	2.2V	5V				
		24p[2]	2.2V	2.2V	5V				
640 × 480		**60p**[2]	0V	5V	0V				
720 × 480	**16:9 Squeeze**	**60p**	0V	5V	5V	EIA-770.2	EIA-770.2	EIA-770.2	−0.300V[3]
	16:9 Squeeze	**30i**	0V	0V	5V				
	16:9 Letterbox	**30i**	0V	0V	2.2V				
	4:3	**30i**	0V	0V	0V				

Notes: 1. 60p, 30i, 30p, and 24p frame rates also include the 59.94p, 29.97i, 29.97p, and 23.976p frame rates.
2. Not part of EIAJ CP-4120 specification, but commonly supported by equipment.
3. Relative to the blanking level.

TABLE 4.13 Common Pro-Video Component Analog Video Formats

Format	Output Signal	Signal Amplitudes (volts)	Notes
SMPTE, EBU N10	Y sync R′ − Y, B′ − Y	+0.700 −0.300 ±0.350	0% setup on Y 100% saturation three wire = (Y + sync), (R′ − Y), (B′ − Y)
525-Line Betacam[1]	Y sync R′ − Y, B′ − Y	+0.714 −0.286 ±0.467	7.5% setup on Y only 100% saturation three wire = (Y + sync), (R′ − Y), (B′ − Y)
625-Line Betacam[1]	Y sync R′ − Y, B′ − Y	+0.700 −0.300 ±0.350	0% setup on Y 100% saturation three wire = (Y + sync), (R′ − Y), (B′ − Y)
525-Line MII[2]	Y sync R′ − Y, B′ − Y	+0.700 −0.300 ±0.324	7.5% setup on Y only 100% saturation three wire = (Y + sync), (R′ − Y), (B′ − Y)
625-Line MII[2]	Y Sync R′ − Y, B′ − Y	+0.700 −0.300 ±0.350	0% setup on Y 100% saturation three wire = (Y + sync), (R′ − Y), (B′ − Y)

Notes: 1. Trademark of Sony Corporation.
2. Trademark of Matsushita Corporation.

Rather than digitize and transmit the blanking intervals, special sequences are inserted into the digital video stream to indicate the *start of active video* (SAV) and *end of active video* (EAV). These sequences indicate when horizontal and vertical blanking is present and which field is being transmitted.

TABLE 4.14 Common Pro-Video RGB Analog Video Formats

Format	Output Signal	Signal Amplitudes (volts)	Notes
SMPTE, EBU N10	G′, B′, R′ sync	+0.700 −0.300	0% setup on G′, B′, and R′ 100% saturation three wire = (G′ + sync), B′, R′
NTSC (setup)	G′, B′, R′ sync	+0.714 −0.286	7.5% setup on G′, B′, and R′ 100% saturation three wire = (G′ + sync), B′, R′
NTSC (no setup)	G′, B′, R′ sync	+0.714 −0.286	0% setup on G′, B′, and R′ 100% saturation three wire = (G′ + sync), B′, R′
MII[1]	G′, B′, R′ sync	+0.700 −0.300	7.5% setup on G′, B′, and R′ 100% saturation three wire = (G′ + sync), B′, R′

Notes: 1. Trademark of Matsushita Corporation.

Pro Video Component Interfaces

Pro video equipment, such as that used within studios, has unique requirements and therefore its own set of digital video interconnect standards. Table 4.20 lists the various pro-video parallel and serial digital interface standards.

Video Timing

As mentioned, the EAV and SAV sequences indicate when horizontal and vertical blanking is present and which field is being transmitted. They also enable

TABLE 4.15 525-Line Betacam 100% Color Bars. Values are relative to the blanking level

		White	Yellow	Cyan	Green	Magenta	Red	Blue	Black
Y	IRE	100	89.5	72.3	61.8	45.7	35.2	18.0	7.5
	mV	714	639	517	441	326	251	129	54
B′ − Y	IRE	0	−65.3	22.0	−43.3	43.3	−22.0	65.3	0
	mV	0	−466	157	−309	309	−157	466	0
R′ − Y	IRE	0	10.6	−65.3	−54.7	54.7	65.3	−10.6	0
	mV	0	76	−466	−391	391	466	−76	0

TABLE 4.16 525-Line Betacam 75% Color Bars. Values are relative to the blanking signal

		White	Yellow	Cyan	Green	Magenta	Red	Blue	Black
Y	IRE	76.9	69.0	56.1	48.2	36.2	28.2	15.4	7.5
	mV	549	492	401	344	258	202	110	54
B′ − Y	IRE	0	−49.0	16.5	−32.5	32.5	−16.5	49.0	0
	mV	0	−350	118	−232	232	−118	350	0
R′ − Y	IRE	0	8.0	−49.0	−41.0	41.0	49.0	−8.0	0
	mV	0	57	−350	−293	293	350	−57	0

TABLE 4.17 525-line MII 100% Color Bars. Values are relative to the blanking signal

		White	Yellow	Cyan	Green	Magenta	Red	Blue	Black
Y	IRE	100	89.5	72.3	61.8	45.7	35.2	18.0	7.5
	mV	700	626	506	433	320	246	126	53
B′ − Y	IRE	0	−46.3	15.6	−30.6	30.6	−15.6	46.3	0
	mV	0	−324	109	−214	214	−109	324	0
R′ − Y	IRE	0	7.5	−46.3	−38.7	38.7	46.3	−7.5	0
	mV	0	53	−324	−271	271	324	−53	0

TABLE 4.18 525-line MII 75% Color Bars. Values are relative to the blanking signal

		White	Yellow	Cyan	Green	Magenta	Red	Blue	Black
Y	**IRE**	76.9	69.0	56.1	48.2	36.2	28.2	15.4	7.5
	mV	538	483	393	338	253	198	108	53
B′ − Y	**IRE**	0	−34.7	11.7	−23.0	23.0	−11.7	34.7	0
	mV	0	−243	82	−161	161	−82	243	0
R′ − Y	**IRE**	0	5.6	−34.7	−29.0	29.0	34.7	−5.6	0
	mV	0	39	−243	−203	203	243	−39	0

TABLE 4.19 VGA Connector Signals

Pin	Function	Signal Level	Impedance
1	red	0.7V	75 ohms
2	green	0.7V	75 ohms
3	blue	0.7V	75 ohms
4	ground		
5	ground		
6	ground - red		
7	ground - green		
8	ground - blue		
9	+5V DC		
10	ground - HSYNC		
11	ground - VSYNC		
12	DDC SDA (data)	≥2.4V	
13	HSYNC (horizontal sync)	≥2.4V	
14	VSYNC (vertical sync)	≥2.4V	
15	DDC SCL (clock)	≥2.4V	

Notes: 1. DDC = Display Data Channel.

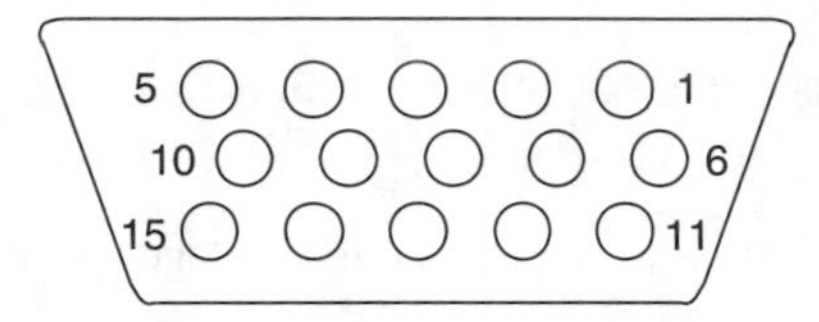

FIGURE 4.11 VGA 15-Pin D-SUB Female Connector.

the transmission of ancillary data such as digital audio, teletext, captioning, etc. during the blanking intervals. The EAV and SAV sequences must have priority over active video data or ancillary data to ensure that correct video timing is always maintained at the receiver. The receiver decodes the EAV and SAV sequences to recover the video timing.

How It Works

The video timing sequence of the encoder is controlled by three timing signals discussed in Chapter 3: H (horizontal blanking), V (vertical blanking), and F (Field 1 or Field 2). A zero-to-one transition of H triggers an EAV sequence while a one-to-zero transition triggers an SAV sequence. F and V are allowed to change only at EAV sequences.

Usually, both 8-bit and 10-bit interfaces are supported, with the 10-bit interface used to transmit 2 bits of fractional video data to minimize cumulative processing errors and to support 10-bit ancillary data.

YCbCr or R′G′B′ data may not use the 10-bit values of 0 × 000–0 × 003 and 0 × 3FC–0 × 3FF, or the 8-bit values of 0 × 00 and 0xFF, since they are used for timing information.

The EAV and SAV sequences are shown in Table 4.21. The status word is defined as:

F = "0" for Field 1 F = "1" for Field 2
V = "1" during vertical blanking
H = "0" at SAV H = "1" at EAV
P3 – P0 = protection bits

$$\begin{aligned} P3 &= V \oplus H \\ P2 &= F \oplus H \\ P1 &= F \oplus V \\ P0 &= F \oplus V \oplus H \end{aligned}$$

where ⊕ represents the exclusive-OR function. These protection bits enable 1- and 2-bit errors to be detected and 1-bit errors to be corrected at the receiver. For most progressive video systems, F is usually a "0" since there is no field information.

TABLE 4.20 Pro-Video Parallel and Serial Digital Interface Standards for Various Component Video Formats (i = interlaced, p = progressive)

Active Resolution (H × V)	Total Resolution[1] (H × V)	Display Aspect Ratio	Frame Rate (Hz)	1 × Y Sample Rate (MHz)	SDTV or HDTV	Digital Parallel Standard	Digital Serial Standard
720 × 480i	858 × 525i	4:3	29.97	13.5	SDTV	BT.656 BT.799 SMPTE 125M	BT.656 BT.799
720 × 480p	858 × 525p	4:3	59.94	27	SDTV	–	BT.1362 SMPTE 294M
720 × 576i	864 × 625i	4:3	25	13.5	SDTV	BT.656 BT.799	BT.656 BT.799
720 × 576p	864 × 625p	4:3	50	27	SDTV	–	BT.1362
960 × 480i	1144 × 525i	16:9	29.97	18	SDTV	BT.1302 BT.1303 SMPTE 267M	BT.1302 BT.1303
960 × 576i	1152 × 625i	16:9	25	18	SDTV	BT.1302 BT.1303	BT.1302 BT.1303
1280 × 720p	1650 × 750p	16:9	59.94	74.176	HDTV	SMPTE 274M	–
1280 × 720p	1650 × 750p	16:9	60	74.25	HDTV	SMPTE 274M	–
1920 × 1080i	2200 × 1125i	16:9	29.97	74.176	HDTV	BT.1120 SMPTE 274M	BT.1120 SMPTE 292M
1920 × 1080i	2200 × 1125i	16:9	30	74.25	HDTV	BT.1120 SMPTE 274M	BT.1120 SMPTE 292M
1920 × 1080p	2200 × 1125p	16:9	59.94	148.35	HDTV	BT.1120 SMPTE 274M	–
1920 × 1080p	2200 × 1125p	16:9	60	148.5	HDTV	BT.1120 SMPTE 274M	–
1920 × 1080i	2376 × 1250i	16:9	25	74.25	HDTV	BT.1120	BT.1120
1920 × 1080p	2376 × 1250p	16:9	50	148.5	HDTV	BT.1120	–

TABLE 4.21 EAV and SAV Sequence

					8-bit Data					10-bit Data
	D9 (MSB)	D8	D7	D6	D5	D4	D3	D2	D1	D0
Preamble	1	1	1	1	1	1	1	1	1	1
	0	0	0	0	0	0	0	0	0	0
	0	0	0	0	0	0	0	0	0	0
Status word	1	F	V	H	P3	P2	P1	P0	0	0

For 4:2:2 YCbCr data, after each SAV sequence, the stream of active data words always begins with a Cb sample, as shown in Figure 4.20. In the multiplexed sequence, the co-sited samples (those that correspond to the same point on the picture) are grouped as Cb, Y, Cr. During blanking intervals, unless ancillary data is present, 10-bit Y or R′G′B′ values should be set to 0 × 040 and 10-bit CbCr values should be set to 0 × 200.

The receiver detects the EAV and SAV sequences by looking for the 8-bit 0xFF 0 × 00 0 × 00 preamble. The status word (optionally error corrected at the receiver, see Table 4.22) is used to recover the H, V, and F timing signals.

Ancillary Data

Ancillary data packets are used to transmit non-video information (such as digital audio, closed captioning, teletext, etc.) during the blanking intervals. A wide variety of ITU-R and SMPTE specifications describe the various ancillary data formats.

During horizontal blanking, ancillary data may be transmitted in the interval between the EAV and SAV sequences. During vertical blanking, ancillary data may be transmitted in the interval between the SAV and EAV sequences. Multiple ancillary packets may be present in a horizontal or vertical blanking interval, but they must be contiguous with each other.

There are two types of ancillary data formats. The older Type 1 format uses a single data ID word to indicate the type of ancillary data; the newer Type 2 format uses two words for the data ID. The general packet format is shown in Table 4.23.

Parallel Interfaces

25-pin Parallel Interface

This interface is used to transfer SDTV resolution 4:2:2 YCbCr data. 8-bit or 10-bit data and a clock are transferred. The individual bits are labeled D0–D9,

TABLE 4.22 SAV and EAV Error Correction at Decoder

Received D5–D2	Received F, V, H (Bits D8–D6)							
	000	**001**	**010**	**011**	**100**	**101**	**110**	**111**
0000	000	000	000	*	000	*	*	111
0001	000	*	*	111	*	111	111	111
0010	000	*	*	011	*	101	*	*
0011	*	*	010	*	100	*	*	111
0100	000	*	*	011	*	*	110	*
0101	*	001	*	*	100	*	*	111
0110	*	011	011	011	100	*	*	011
0111	100	*	*	011	100	100	100	*
1000	000	*	*	*	*	101	110	*
1001	*	001	010	*	*	*	*	111
1010	*	101	010	*	101	101	*	101
1011	010	*	010	010	*	101	010	*
1100	*	001	110	*	110	*	110	110
1101	001	001	*	001	*	001	110	*
1110	*	*	*	011	*	101	110	*
1111	*	001	010	*	100	*	*	*

*Notes: * = uncorrectable error.*

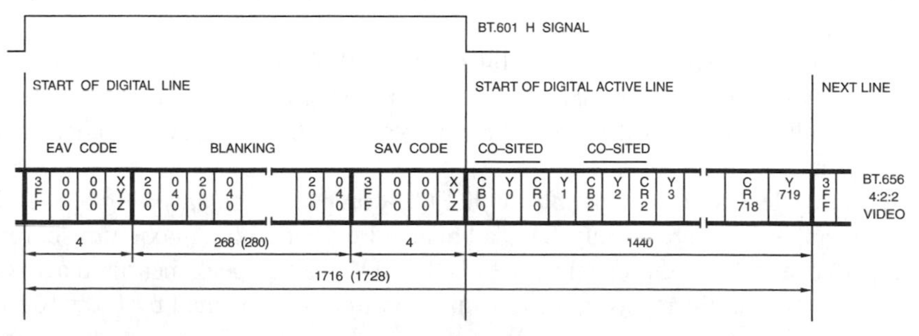

FIGURE 4.12 BT.656 Parallel Interface Data For One Scan Line. 480i; 4:2:2 YCbCr; 720 active samples per line; 27 MHz clock; 10-bit system. The values for 576i systems are shown in parentheses.

TABLE 4.23 Ancillary Data Packet General Format

	8-bit Data								10-bit Data	
	D9 (MSB)	D8	D7	D6	D5	D4	D3	D2	D1	D0
ancillary data flag (ADF)	0	0	0	0	0	0	0	0	0	0
	1	1	1	1	1	1	1	1	1	1
	1	1	1	1	1	1	1	1	1	1
data ID (DID)	$\overline{D8}$	even parity			value of 0000 0000 to 1111 1111					
data block number or SDID	$\overline{D8}$	even parity			value of 0000 0000 to 1111 1111					
Data count (DC)	$\overline{D8}$	even parity			value of 0000 0000 to 1111 1111					
user data word 0				value of 00 0000 0100 to 11 1111 1011						
						:				
user data word N				value of 00 0000 0100 to 11 1111 1011						
checksum	$\overline{D8}$	sum of D0–D8 of data ID through last user data word. Preset to all zeros; carry is ignored.								

with D9 being the most significant bit. The pin allocations for the signals are shown in Table 4.24.

Y has a nominal 10-bit range of 0 × 040–0 × 3AC. Values less than 0 × 040 or greater than 0 × 3AC may be present due to processing. During blanking, Y data should have a value of 040H, unless other information is present.

Cb and Cr have a nominal 10-bit range of 0 × 040–0 × 3C0. Values less than 0 × 040 or greater than 0 × 3C0 may be present due to processing. During blanking, CbCr data should have a value of 0 × 200, unless other data is present.

Signal levels are compatible with ECL-compatible balanced drivers and receivers. The generator must have a balanced output with a maximum source impedance of 110?; the signal must be 0.8–2.0 V peak-to-peak measured across a 110-? load. At the receiver, the transmission line is terminated by 110±10?

27 MHz Parallel Interface: This BT.656 and SMPTE 125M interface is used for 480i and 576i systems with an aspect ratio of 4:3. Y and multiplexed CbCr

TABLE 4.24 25-Pin Parallel Interface Connector Pin Assignments. For 8-bit interfaces, D9–D2 are used

Pin	Signal	Pin	Signal
1	clock	14	clock–
2	system ground A	15	system ground B
3	D9	16	D9–
4	D8	17	D8–
5	D7	18	D7–
6	D6	19	D6–
7	D5	20	D5–
8	D4	21	D4–
9	D3	22	D3–
10	D2	23	D2–
11	D1	24	D1–
12	D0	25	D0–
13	cable shield		

information at a sample rate of 13.5 MHz are multiplexed into a single 8-bit or 10-bit data stream, at a clock rate of 27 MHz.

The 27 MHz clock signal has a clock pulse width of 18.5±3 ns. The positive transition of the clock signal occurs midway between data transitions with a tolerance of ±3 ns (as shown in Figure 4.13).

To permit reliable operation at interconnect lengths of 50–200 meters, the receiver must use frequency equalization, with typical characteristics shown in Figure 4.14. This example enables operation with a range of cable lengths down to zero.

36 MHz Parallel Interface: This BT.1302 and SMPTE 267M interface is used for 480i and 576i systems with an aspect ratio of 16:9. Y and multiplexed CbCr information at a sample rate of 18 MHz are multiplexed into a single 8-bit or 10-bit data stream, at a clock rate of 36 MHz.

The 36 MHz clock signal has a clock pulse width of 13.9±2 ns. The positive transition of the clock signal occurs midway between data transitions with a tolerance of ±2 ns (as shown in Figure 4.15).

To permit reliable operation at interconnect lengths of 40–160 meters, the receiver must use frequency equalization, with typical characteristics shown in Figure 4.14.

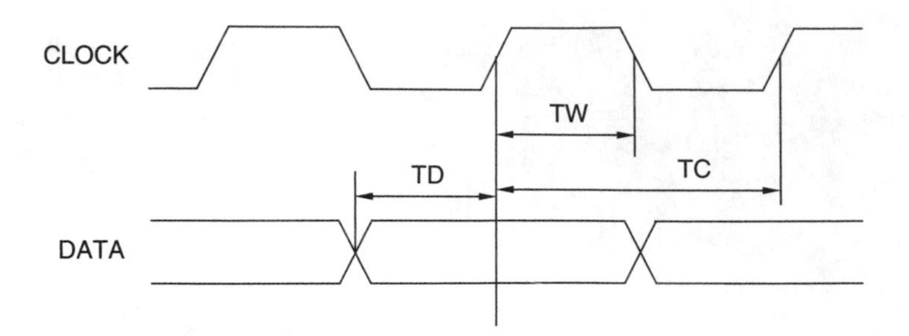

TW = 18.5 ± 3 NS
TC = 37 NS
TD = 18.5 ± 3 NS

FIGURE 4.13 25-Pin 27 MHz Parallel Interface Waveforms.

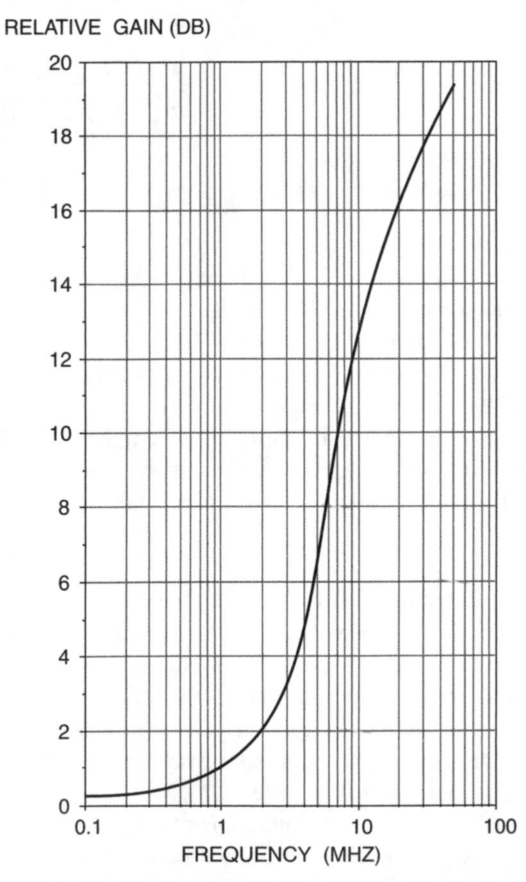

FIGURE 4.14 Example Line Receiver Equalization Characteristics for Small Signals.

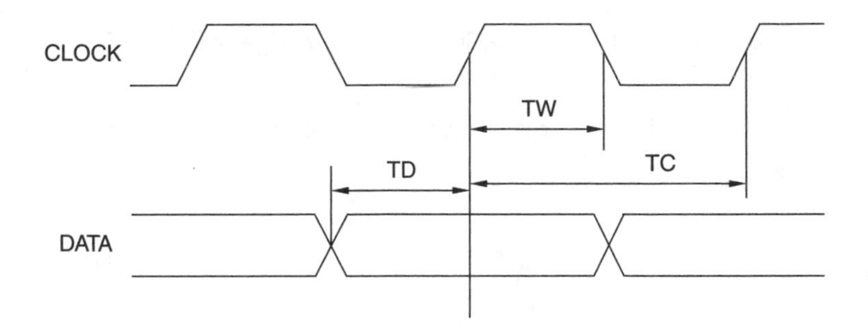

TW = 13.9 ± 2 NS

TC = 27.8 NS

TD = 13.9 ± 2 NS

FIGURE 4.15 25-Pin 36 MHz Parallel Interface Waveforms.

93-pin Parallel Interface

This interface is used to transfer HDTV resolution R′G′B′ data, 4:2:2 YCbCr data, or 4:2:2:4 YCbCrK data. The pin allocations for the signals are shown in Table 4.25. The most significant bits are R9, G9, and B9.

When transferring 4:2:2 YCbCr data, the green channel carries Y information and the red channel carries multiplexed CbCr information.

When transferring 4:2:2:4 YCbCrK data, the green channel carries Y information, the red channel carries multiplexed CbCr information, and the blue channel carries K (alpha keying) information.

Y has a nominal 10-bit range of 0 × 040−0 × 3AC. Values less than 040H or greater than 0 × 3AC may be present due to processing. During blanking, Y data should have a value of 0 × 040, unless other information is present.

Cb and Cr have a nominal 10-bit range of 0 × 040–0 × 3C0. Values less than 0 × 040 or greater than 0 × 3C0 may be present due to processing. During blanking, CbCr data should have a value of 0 × 200, unless other information is present.

R′G′B′ and K have a nominal 10-bit range of 0 × 040–0 × 3AC. Values less than 0 × 040 or greater than 0 × 3AC may be present due to processing. During blanking, R′G′B′ data should have a value of 0 × 040, unless other information is present.

Signal levels are compatible with ECL-compatible balanced drivers and receivers. The generator must have a balanced output with a maximum source impedance of 110Ω; the signal must be 0.6–2.0 V peak-to-peak measured across a 110Ω load. At the receiver, the transmission line must be terminated by 110±10.

74.25 and 74.176 MHz Parallel Interface: This ITU-R BT.1120 and SMPTE 274M interface is primarily used for HDTV systems. The 74.25 or 74.176 MHz

TABLE 4.25 **93-Pin Parallel Interface Connector Pin Assignments. For 8-bit interfaces, bits 9–2 are used**

Pin	Signal	Pin	Signal	Pin	Signal	Pin	Signal
1	clock	**26**	GND	**51**	B2	**76**	GND
2	G9	**27**	GND	**52**	B1	**77**	GND
3	G8	**28**	GND	**53**	B0	**78**	GND
4	G7	**29**	GND	**54**	R9	**79**	B4–
5	G6	**30**	GND	**55**	R8	**80**	B3–
6	G5	**31**	GND	**56**	R7	**81**	B2–
7	G4	**32**	GND	**57**	R6	**82**	B1–
8	G3	**33**	clock–	**58**	R5	**83**	B0–
9	G2	**34**	G9–	**59**	R4	**84**	R9–
10	G1	**35**	G8–	**60**	R3	**85**	R8–
11	G0	**36**	G7–	**61**	R2	**86**	R7–
12	B9	**37**	G6–	**62**	R1	**87**	R6–
13	B8	**38**	G5–	**63**	R0	**88**	R5–
14	B7	**39**	G4–	**64**	GND	**89**	R4–
15	B6	**40**	G3–	**65**	GND	**90**	R3–
16	B5	**41**	G2–	**66**	GND	**91**	R2–
17	GND	**42**	G1–	**67**	GND	**92**	R1–
18	GND	**43**	G0–	**68**	GND	**93**	R0–
19	GND	**44**	B9–	**69**	GND		
20	GND	**45**	B8–	**70**	GND		
21	GND	**46**	B7–	**71**	GND		
22	GND	**47**	B6–	**72**	GND		
23	GND	**48**	B5–	**73**	GND		
24	GND	**49**	B4	**74**	GND		
25	GND	**50**	B3	**75**	GND		

(74.25/1.001) clock signal has a clock pulse width of 6.73±1.48 ns. The positive transition of the clock signal occurs midway between data transitions with a tolerance of ±1 ns (as shown in Figure 4.16).

To permit reliable operation at interconnect lengths greater than 20 meters, the receiver must use frequency equalization.

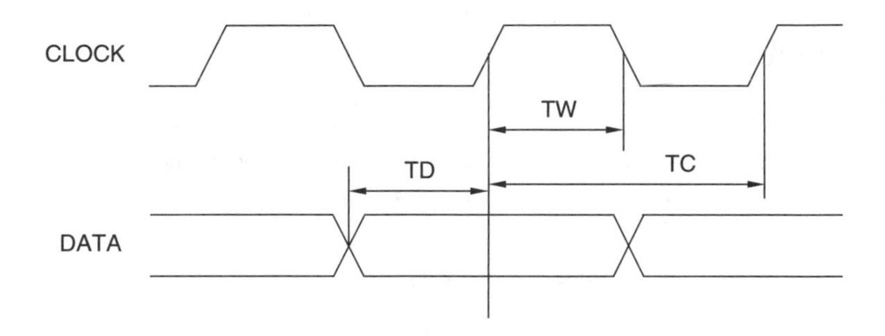

FIGURE 4.16 93-Pin 74.25 and 74.176 MHz Parallel Interface Waveforms.

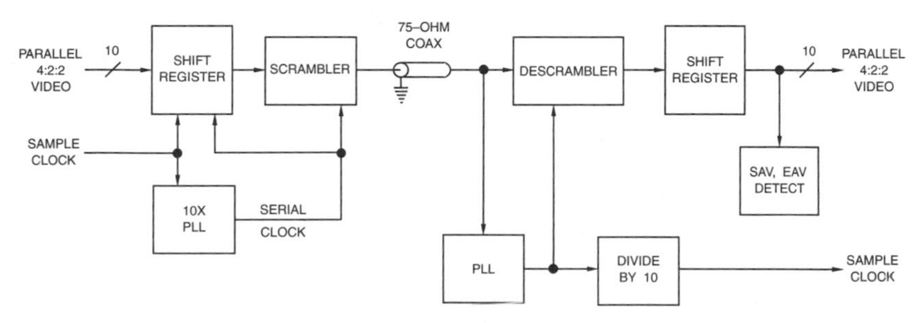

FIGURE 4.17 Serial Interface Block Diagram.

148.5 and 148.35 MHz Parallel Interface: This BT.1120 and SMPTE 274M interface is used for HDTV systems. The 148.5 or 148.35 MHz (148.5/1.001) clock signal has a clock pulse width of 3.37±0.74 ns. The positive transition of the clock signal occurs midway between data transitions with a tolerance of ±0.5 ns (similar to Figure 4.16). To permit reliable operation at interconnect lengths greater than 14 meters, the receiver must use frequency equalization.

Serial Interfaces

The parallel formats can be converted to a serial format (Figure 4.17), allowing data to be transmitted using a 75 Ω coaxial cable or optical fiber.

For cable interconnect, the generator has an unbalanced output with a source impedance of 75 Ω; the signal must be 0.8 V ±10% peak-to-peak measured across a 75 Ω load. The receiver has an input impedance of 75 Ω.

How It Works

In an 8-bit environment, before serialization, the 0 × 00 and 0xFF codes during EAV and SAV are expanded to 10-bit values of 0 × 000 and 0 × 3FF, respectively. All other 8-bit data is appended with two least significant "0" bits before serialization.

The 10 bits of data are serialized (LSB first) and processed using a scrambled and polarity-free NRZI algorithm:

$$G(x) = (x^9 + x^4 + 1)(x + 1)$$

The input signal to the scrambler (Figure 4.18) uses positive logic (the highest voltage represents a logical one; lowest voltage represents a logical zero). The formatted serial data is output at the 10 × sample clock rate. Since the parallel clock may contain large amounts of jitter, deriving the 10 × sample clock directly from an unfiltered parallel clock may result in excessive signal jitter.

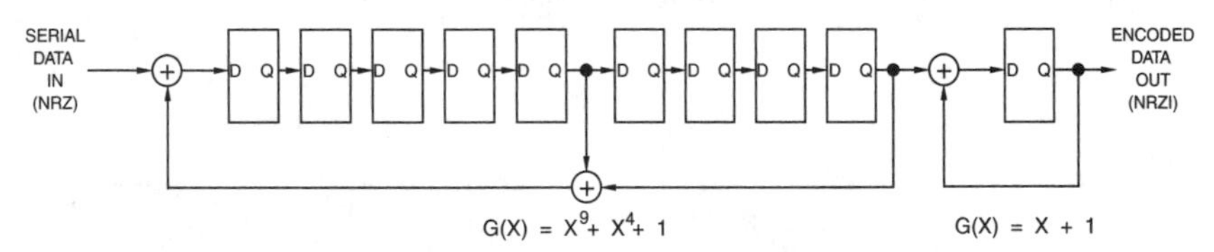

FIGURE 4.18 Typical Scrambler Circuit.

At the receiver, phase-lock synchronization is done by detecting the EAV and SAV sequences. The PLL is continuously adjusted slightly each scan line to ensure that these patterns are detected and to avoid bit slippage. The recovered 10× sample clock is divided by ten to generate the sample clock, although care must be taken not to mask word-related jitter components. The serial data is low- and high-frequency equalized, inverse scrambling performed (Figure 4.19), and deserialized.

Pro-Video Composite Interfaces

Digital composite video is essentially a digital version of a composite analog (M) NTSC or (B, D, G, H, I) PAL video signal. The sample clock rate is four times FSC: about 14.32 MHz for (M) NTSC and about 17.73 MHz for (B, D, G, H, I) PAL.

Usually, both 8-bit and 10-bit interfaces are supported, with the 10-bit interface used to transmit 2 bits of fractional video data to minimize cumulative processing errors and to support 10-bit ancillary data.

Table 4.26 lists the digital composite levels. Video data may not use the 10-bit values of 0 × 000–0 × 003 and 0 × 3FC−0 × 3FF, or the 8-bit values of 0 × 00 and 0xFF, since they are used for timing information.

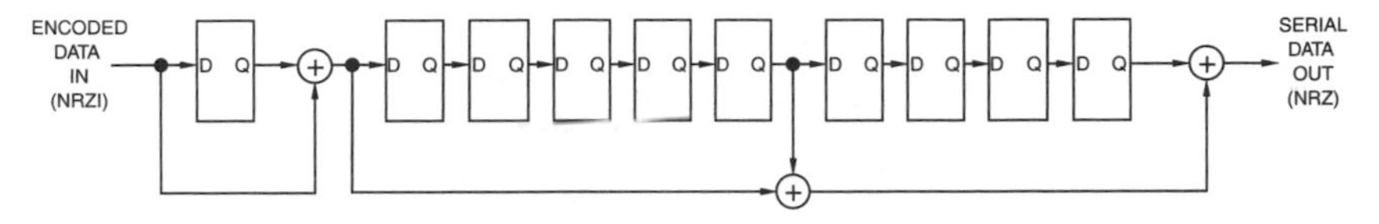

FIGURE 4.19 Typical Descrambler Circuit.

TABLE 4.26 10-Bit Video Levels for Digital Composite Video Signals

Video Level	(M) NTSC	(B, D, G, H, I) PAL
peak chroma	972	1040 (limited to 1023)
White	800	844
peak burst	352	380
Black	280	256
Blank	240	256
peak burst	128	128
peak chroma	104	128
Sync	16	4

NTSC Video Timing

There are 910 total samples per scan line, as shown in Figure 4.20. Horizontal count 0 corresponds to the start of active video, and a horizontal count of 768 corresponds to the start of horizontal blanking.

Sampling is along the ±I and ±Q axes (33°, 123°, 213°, and 303°). The sampling phase at horizontal count 0 of line 10, Field 1 is on the +V axis (123°).

The sync edge values, and the horizontal counts at which they occur, are defined as shown in Figure 4.21 and Tables 4.27–4.29. 8-bit values for one color burst cycle are 45, 83, 75, and 37. The burst envelope starts at horizontal count 857, and lasts for 43 clock cycles, as shown in Table 4.27. Note that the peak amplitudes of the burst are not sampled.

To maintain zero SCH phase, horizontal count 784 occurs 25.6 ns (33° of the subcarrier phase) before the 50% point of the falling edge of horizontal sync, and horizontal count 785 occurs 44.2 ns (57° of the subcarrier phase) after the 50% point of the falling edge of horizontal sync.

PAL Video Timing

There are 1135 total samples per line, except for two lines per frame which have 1137 samples per line, making a total of 709,379 samples per frame. Figure 4.22 illustrates the typical line timing. Horizontal count 0 corresponds

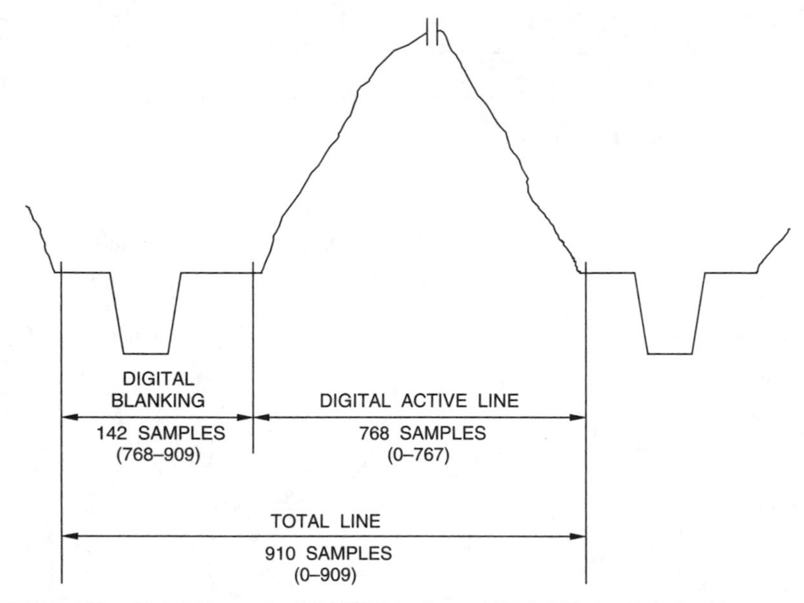

FIGURE 4.20 Digital Composite (M) NTSC Analog and Digital Timing Relationship.

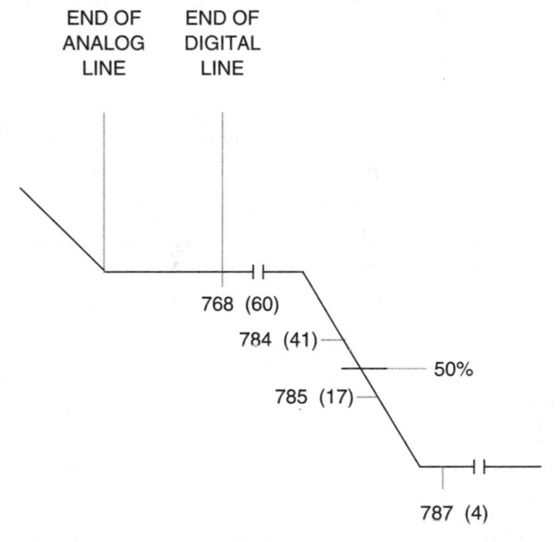

FIGURE 4.21 Digital Composite (M) NTSC Sync Timing. The horizontal counts are shown with the corresponding 8-bit sample values in parentheses.

to the start of active video, and a horizontal count of 948 corresponds to the start of horizontal blanking.

Sampling is along the ±U and ±V axes (0°, 90°, 180°, and 270°), with the sampling phase at horizontal count 0 of line 1, Field 1 on the +V axis (90°).

TABLE 4.27A Digital Values During the Horizontal Blanking Intervals for Digital Composite (M) NTSC Video Signals

	8-bit Hex Value		10-bit Hex Value	
Sample	**Fields 1, 3**	**Fields 2, 4**	**Fields 1, 3**	**Fields 2,4**
768–782	3C	3C	0F0	0F0
783	3A	3A	0E9	0E9
784	29	29	0A4	0A4
785	11	11	044	044
786	04	04	011	011
787–849	04	04	010	010
850	06	06	017	017
851	17	17	05C	05C
852	2F	2F	0BC	0BC
853	3C	3C	0EF	0EF
854–856	3C	3C	0F0	0F0
857	3C	3C	0F0	0F0
858	3D	3B	0F4	0EC
859	37	41	0DC	104
860	36	42	0D6	10A
861	4B	2D	12C	0B4
862	49	2F	123	0BD
863	25	53	096	14A
864	2D	4B	0B3	12D
865	53	25	14E	092
866	4B	2D	12D	0B3
867	25	53	092	14E
868	2D	4B	0B3	12D
869	53	25	14E	092
870	4B	2D	12D	0B3
871	25	53	092	14E
872	2D	4B	0B3	12D
873	53	25	14E	092

TABLE 4.27B Digital Values During the Horizontal Blanking Intervals for Digital Composite (M) NTSC Video Signals

	8-bit Hex Value		10-bit Hex Value	
Sample	**Fields 1, 3**	**Fields 2, 4**	**Fields 1, 3**	**Fields 2,4**
874	4B	2D	12D	0B3
875	25	53	092	14E
876	2D	4B	0B3	12D
877	53	25	14E	092
878	4B	2D	12D	0B3
879	25	53	092	14E
880	2D	4B	0B3	12D
881	53	25	14E	092
882	4B	2D	12D	0B3
883	25	53	092	14E
884	2D	4B	0B3	12D
885	53	25	14E	092
886	4B	2D	12D	0B3
887	25	53	092	14E
888	2D	4B	0B3	12D
889	53	25	14E	092
890	4B	2D	12D	0B3
891	25	53	092	14E
892	2D	4B	0B3	12D
893	53	25	14E	092
894	4A	2E	129	0B7
895	2A	4E	0A6	13A
896	33	45	0CD	113
897	44	34	112	0CE
898	3F	39	0FA	0E6
899	3B	3D	0EC	0F4
900–909	3C	3C	0F0	0F0

TABLE 4.28 Equalizing Pulse Values During the Vertical Blanking Intervals for Digital Composite (M) NTSC Video Signals

Fields 1, 3			Fields 2,4		
Sample	8-bit Hex Value	10-bit Hex Value	Sample	8-bit Hex Value	10-bit Hex Value
768–782	3C	0F0	**313–327**	3C	0F0
783	3A	0E9	**328**	3A	0E9
784	29	0A4	**329**	29	0A4
785	11	044	**330**	11	044
786	04	011	**331**	04	011
787–815	04	010	**332–360**	04	010
816	06	017	**361**	06	017
817	17	05C	**362**	17	05C
818	2F	0BC	**363**	2F	0BC
819	3C	0EF	**364**	3C	0EF
820–327	3C	0F0	**365–782**	3C	0F0
328	3A	0E9	**783**	3A	0E9
329	29	0A4	**784**	29	0A4
330	11	044	**785**	11	044
331	04	011	**786**	04	011
332–360	04	010	**787–815**	04	010
361	06	017	**816**	06	017
362	17	05C	**817**	17	05C
363	2F	0BC	**818**	2F	0BC
364	3C	0EF	**819**	3C	0EF
365–782	3C	0F0	**820–327**	3C	0F0

8-bit color burst values are 95, 64, 32, and 64, continuously repeated. The swinging burst causes the peak burst (32 and 95) and zero burst (64) samples to change places. The burst envelope starts at horizontal count 1058, and lasts for 40 clock cycles.

Sampling is not H-coherent as with (M) NTSC, so the position of the sync pulses changes from line to line. Zero SCH phase is defined when alternate burst samples have a value of 64.

TABLE 4.29 Serration Pulse Values During the Vertical Blanking Intervals for Digital Composite (M) NTSC Video Signals

	Fields 1, 3			Fields 2,4	
Sample	8-bit Hex Value	10-bit Hex Value	Sample	8-bit Hex Value	10-bit Hex Value
782	3C	0F0	**327**	3C	0F0
783	3A	0E9	**328**	3A	0E9
784	29	0A4	**329**	29	0A4
785	11	044	**330**	11	044
786	04	011	**331**	04	011
787–260	04	010	**332–715**	04	010
261	06	017	**716**	06	017
262	17	05C	**717**	17	05C
263	2F	0BC	**718**	2F	0BC
264	3C	0EF	**719**	3C	0EF
265–327	3C	0F0	**720–782**	3C	0F0
328	3A	0E9	**783**	3A	0E9
329	29	0A4	**784**	29	0A4
330	11	044	**785**	11	044
331	04	011	**786**	04	011
332–715	04	010	**787–260**	04	010
716	06	017	**261**	06	017
717	17	05C	**262**	17	05C
718	2F	0BC	**263**	2F	0BC
719	3C	0EF	**264**	3C	0EF
720–782	3C	0F0	**265–327**	3C	0F0

Ancillary Data

Ancillary data packets are used to transmit information (such as digital audio, closed captioning, and teletext data) during the blanking intervals. ITU-R BT.1364 and SMPTE 291M describe the ancillary data formats.

The ancillary data formats are the same as for digital component video, discussed earlier in this chapter. However, instead of a 3-word preamble, a one-word ancillary data flag is used, with a 10-bit value of 3FCH. There may be

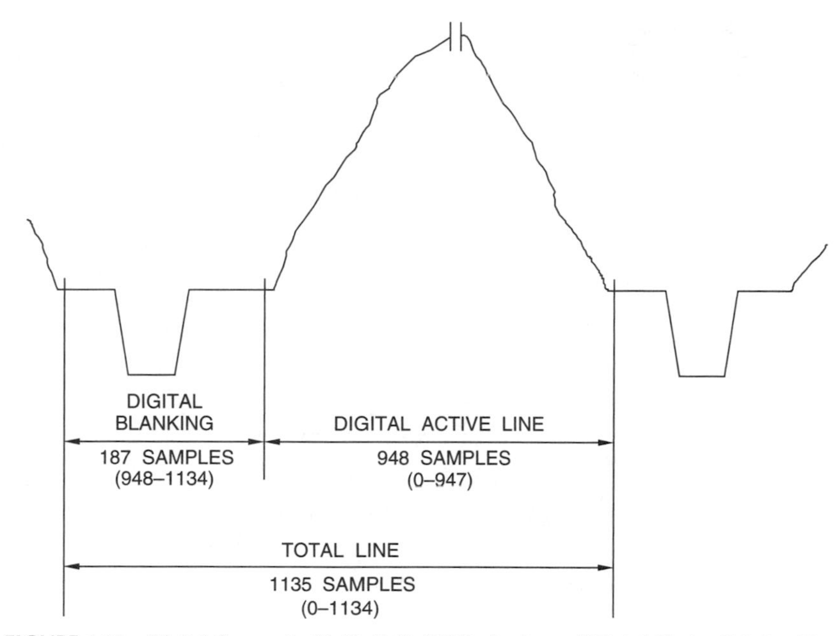

FIGURE 4.22 Digital Composite (B, D, G, H, I) PAL Analog and Digital Timing Relationship.

multiple ancillary data flags following the TRS-ID, with each flag identifying the beginning of another ancillary packet.

Ancillary data may be present within the following word number boundaries (see Figures 4.23 through 4.28).

NTSC	PAL	
795–849	972–1035	horizontal sync period
795–815	972–994	equalizing pulse periods
340–360	404–426	
795–260	972–302	vertical sync periods
340–715	404–869	

User data may not use the 10-bit values of 0 × 000–0 × 003 and 0 × 3FC–0 × 3FF, or the 8-bit values of 0 × 00 and 0xFF, since they are used for timing information.

Parallel Interface

The SMPTE 244M 25-pin parallel interface is based on that used for 27 MHz 4:2:2 digital component video (Table 4.24), except for the timing differences.

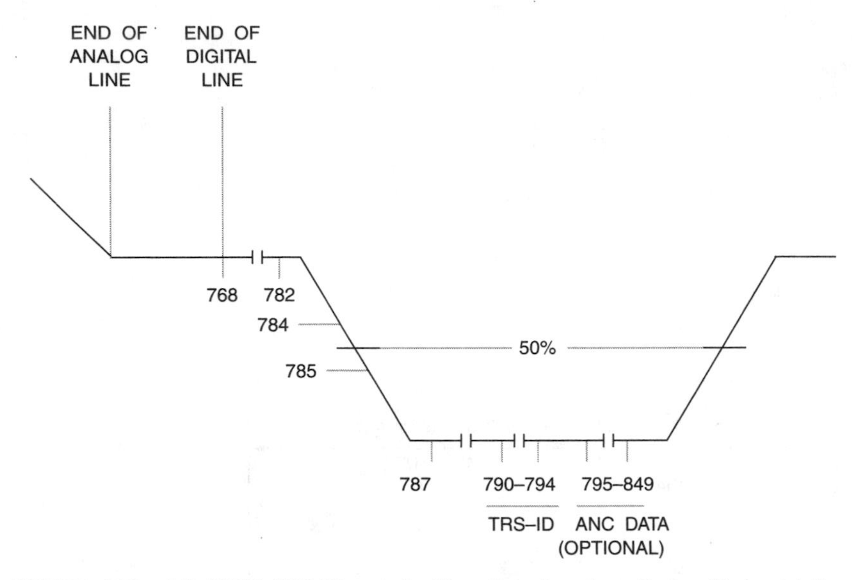

FIGURE 4.23 (M) NTSC TRS-ID and Ancillary Data Locations During Horizontal Sync Intervals.

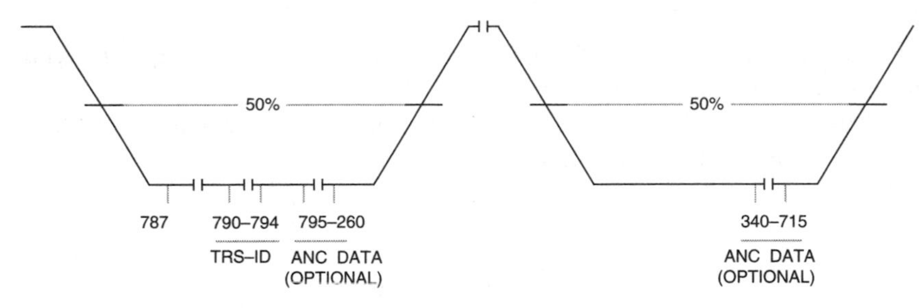

FIGURE 4.24 (M) NTSC TRS-ID and Ancillary Data Locations During Vertical Sync Intervals.

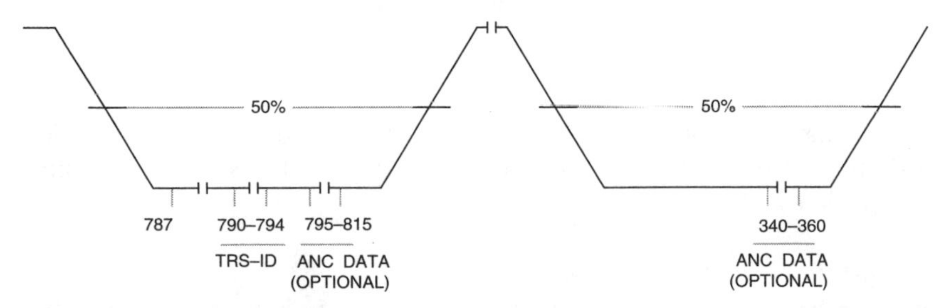

FIGURE 4.25 (M) NTSC TRS-ID and Ancillary Data Locations During Equalizing Pulse Intervals.

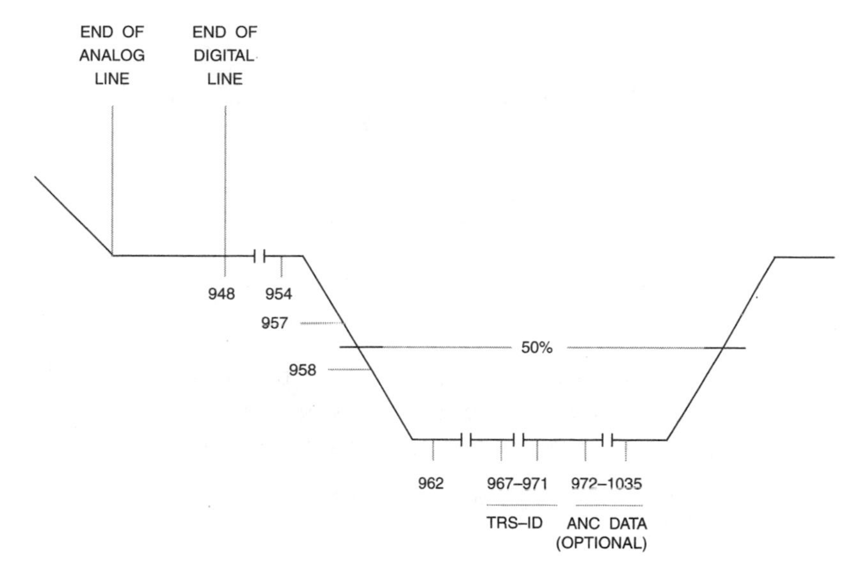

FIGURE 4.26 (B, D, G, H, I) PAL TRS-ID and Ancillary Data Locations During Horizontal Sync Intervals.

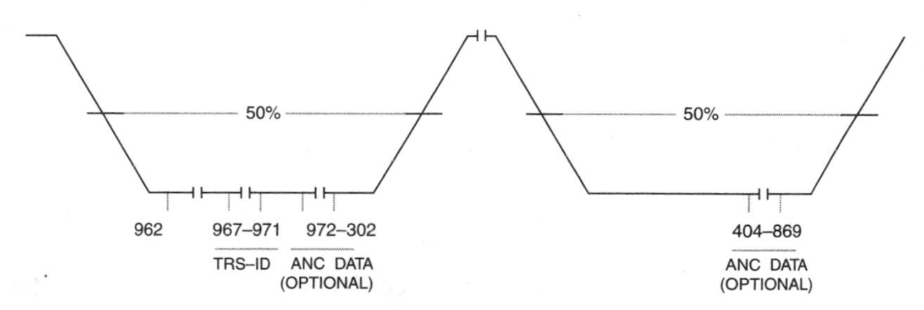

FIGURE 4.27 (B, D, G, H, I) PAL TRS-ID and Ancillary Data Locations During Vertical Sync Intervals.

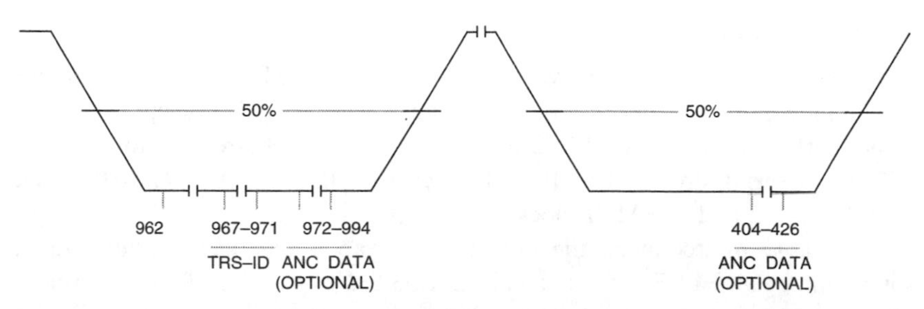

FIGURE 4.28 (B, D, G, H, I) PAL TRS-ID and Ancillary Data Locations During Equalizing Pulse Intervals.

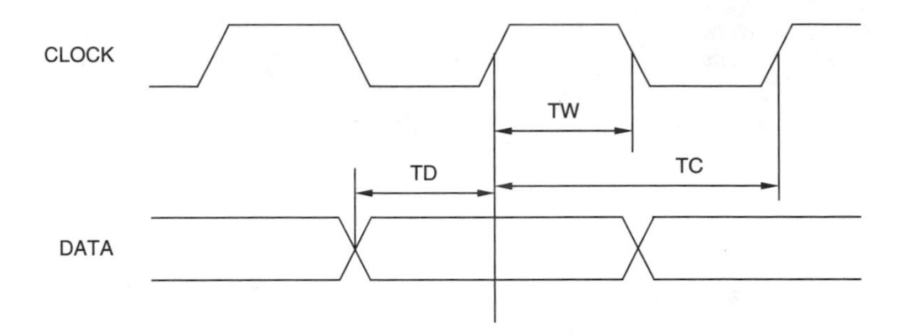

TW = 35 ± 5 NS (M) NTSC; 28 ± 5 NS (B, D, G, H, I) PAL

TC = 69.84 NS (M) NTSC; 56.39 NS (B, D, G, H, I) PAL

TD = 35 ± 5 NS (M) NTSC; 28 ± 5 NS (B, D, G, H, I) PAL

FIGURE 4.29 Digital Composite Video Parallel Interface Waveforms.

This interface is used to transfer SDTV resolution digital composite data. 8-bit or 10-bit data and a 4 × FSC clock are transferred.

Signal levels are compatible with ECL-compatible balanced drivers and receivers. The generator must have a balanced output with a maximum source impedance of 110 Ω; the signal must be 0.8–2.0 V peak-to-peak measured across a 110 Ω load. At the receiver, the transmission line must be terminated by $110 \pm 10\,\Omega$.

The clock signal is a 4 × FSC square wave, with a clock pulse width of 35 ±5 ns for (M) NTSC or 28 ±5 ns for (B, D, G, H, I) PAL. The positive transition of the clock signal occurs midway between data transitions with a tolerance of ±5 ns (as shown in Figure 4.29).

To permit reliable operation at interconnect lengths of 50–200 meters, the receiver must use frequency equalization, with typical characteristics shown in Figure 4.14. This example enables operation with a range of cable lengths down to zero.

Serial Interface

The parallel format can be converted to a SMPTE 259M serial format (Figure 4.30), allowing data to be transmitted using a coaxial cable (or optical fiber). This interface converts the 14.32 or 17.73 MHz parallel stream into a 143 or 177 Mbps serial stream. The 10× PLL generates the 143 or 177 MHz clock from the 14.32 or 17.73 MHz clock signal.

For cable interconnect, the generator has an unbalanced output with a source impedance of 75 ?; the signal must be 0.8 V ±10% peak-to-peak measured across a 75 Ω load. The receiver has an input impedance of 75 Ω.

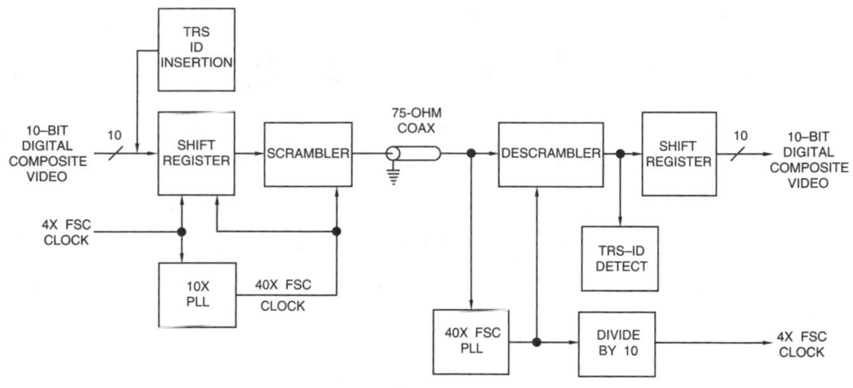

FIGURE 4.30 Serial Interface Block Diagram.

How It Works

The 10 bits of data are serialized (LSB first) and processed using a scrambled and polarity-free NRZI algorithm:

$$G(x) = (x^9 + x^4 + 1)(x + 1)$$

This algorithm is the same as used for digital component video discussed earlier. In an 8-bit environment, 8-bit data is appended with two least significant "0" bits before serialization.

The input signal to the scrambler (Figure 4.18) uses positive logic (the highest voltage represents a logical one; lowest voltage represents a logical zero). The formatted serial data is output at the 40× FSC rate.

At the receiver, phase-lock synchronization is done by detecting the TRS-ID sequences. The PLL is continuously adjusted slightly each scan line to ensure that these patterns are detected and to avoid bit slippage. The recovered 10× clock is divided by ten to generate the 4× FSC sample clock. The serial data is low- and high-frequency equalized, inverse scrambling performed (Figure 4.19), and deserialized.

Pro-Video Transport Interfaces

Serial Data Transport Interface (SDTI)

SMPTE 305M and ITU-R BT.1381 define a Serial Data Transport Interface (SDTI) that enables transferring data between equipment. The physical layer uses the 270 or 360 Mbps BT.656, BT.1302, and SMPTE 259M digital component video serial interface. Figure 4.31 illustrates the signal format.

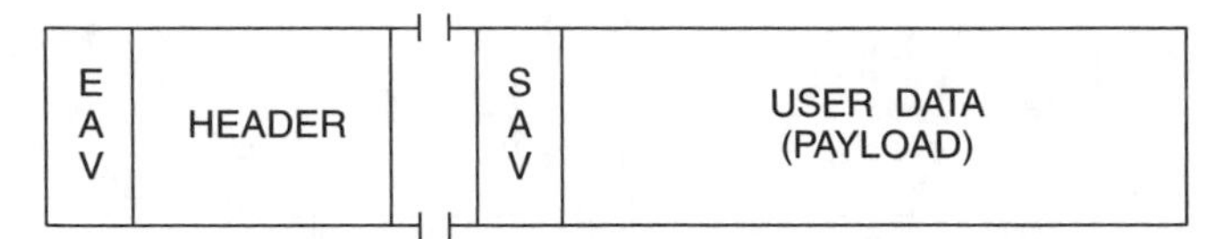

FIGURE 4.31 SDTI Signal Format.

TABLE 4.30 DVI-D Connector Signal Assignments

Pin	Signal	Pin	Signal	Pin	Signal
1	D2−	9	D1−	17	D0−
2	D2	10	D1	18	D0
3	shield	11	shield	19	shield
4	D4−	12	D3−	20	D5−
5	D4	13	D3	21	D5
6	DDC SCL	14	+5V	22	shield
7	DDC SDA	15	ground	23	CLK
8	reserved	16	Hot Plug Detect	24	CLK−

A 53-word header is inserted immediately after the EAV sequence, specifying the source, destination, and data format. Table 4.30 illustrates the header contents.

The payload data is defined within BT.1381 and by other application-specific standards such as SMPTE 326M. It may consist of MPEG-2 program or transport streams, DV streams, etc., and uses either 8-bit words plus even parity and D8 , or 9-bit words plus D8 .

High Data-Rate Serial Data Transport Interface (HD-SDTI)

SMPTE 348M and ITU-R BT.1577 define a High Data-Rate Serial Data Transport Interface (HD-SDTI) that enables transferring data between equipment. The physical layer uses the 1.485 (or 1.485/1.001) Gbps SMPTE 292M digital component video serial interface.

Figure 4.32 illustrates the signal format. Two data channels are multiplexed onto the single HD-SDTI stream such that one 74.25 (or 74.25/1.001) MHz data stream occupies the Y data space and the other 74.25 (or 74.25/1.001) MHz data stream occupies the CbCr data space.

A 49-word header is inserted immediately after the line number CRC data, specifying the source, destination, and data format.

The payload data is defined by other application-specific standards. It may consist of MPEG-2 program or transport streams, DV streams, etc., and uses either 8-bit words plus even parity and D8 , or 9-bit words plus D8 .

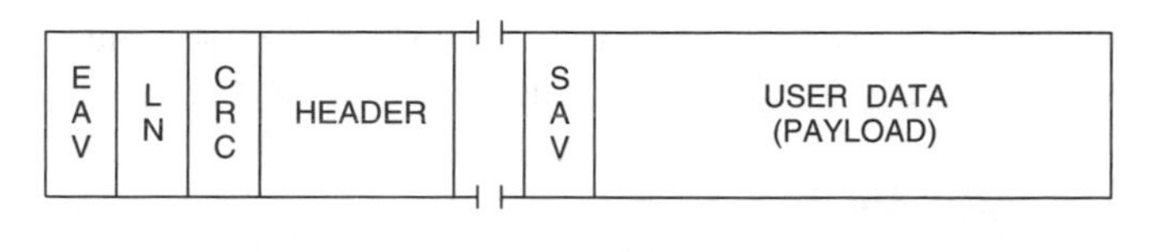

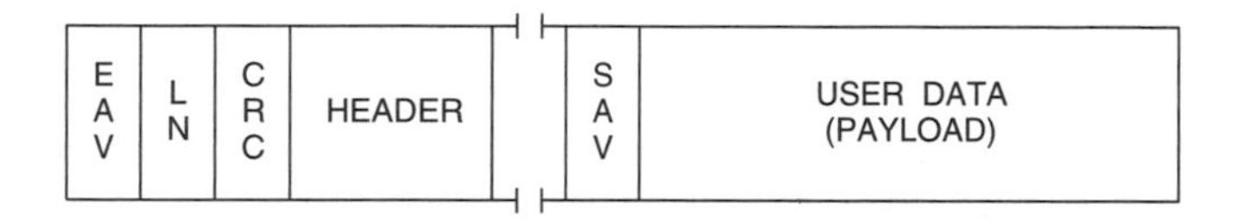

FIGURE 4.32 HD-SDTI Signal Format. LN = line number (two 10-bit words), CRC = line number CRC (two 10-bit words).

IC Component Interfaces

Many solutions for transferring digital video between chips are derived from the pro-video interconnect standards. Chips for the pro-video market typically support 10 or 12 bits of data per video component, while chips for the consumer market typically use 8 bits of data per video component. BT.601 and BT.656 are the most popular interfaces.

YCbCr Values: 8-bit Data

Y has a nominal range of 0 × 10–0xEB. Values less than 10H or greater than 0xEBH may be present due to processing. Cb and Cr have a nominal range of 0 × 10–0xF0. Values less than 0 × 10 or greater than 0xF0 may be present due to processing. YCbCr data may not use the values of 00H and FFH since those values may be used for timing information.

During blanking, Y data should have a value of 0 × 10 and CbCr data should have a value of 0 × 80, unless other information is present.

YCbCr Values: 10-bit Data

For higher accuracy, pro-video solutions typically use 10-bit YCbCr data. Y has a nominal range of 0 × 040–0 × 3AC. Values less than 0 × 040 or greater than 0 × 3AC may be present due to processing. Cb and Cr have a nominal range of 040H–3C0H. Values less than 0 × 040 or greater than 0 × 3C0 may be present due to processing. The values 0 × 000–0 × 003 and 0 × 3FC–0× 3FF may not be used to avoid timing contention with 8-bit systems.

During blanking, Y data should have a value of 0 × 040 and CbCr data should have a value of 0 × 200, unless other information is present.

RGB Values: 8-bit Data

Consumer solutions typically use 8-bit R′G′B′ data, with a range of 0 × 10–0xEB (note that PCs typically use a range of 0 × 00–0xFF). Values less than 0 × 10 or greater than 0xEB may be present due to processing.

During blanking, R′G′B′ data should have a value of 0 × 10, unless other information is present.

RGB Values: 10-bit Data

For higher accuracy, pro-video solutions typically use 10-bit R′G′B′ data, with a nominal range of 0 × 040–0 × 3AC. Values less than 0 × 040 or greater than 0 × 3ACH may be present due to processing. The values 0 × 000–0 × 003 and 0 × 3FC–0 × 3FF may not be used to avoid timing contention with 8-bit systems.

During blanking, R′G′B′ data should have a value of 0 × 040, unless other data is present.

BT.601 Video Interface

The BT.601 video interface has been used for years, with the control signal names and timing reflecting the video standard. Supported active resolutions and sample clock rates are dependent on the video standard and aspect ratio.

Devices usually support multiple data formats to simplify using them in a wide variety of applications.

Video Module Interface (VMI)

VMI (Video Module Interface) was developed in cooperation with several multimedia IC manufacturers. The goal was to standardize the video interfaces between devices such as MPEG decoders, NTSC/PAL decoders, and graphics chips.

Video Data Formats

The VMI specification specifies an 8-bit 4:2:2 YCbCr data format as shown in Figure 4.33. Many devices also support the other YCbCr and R′G′B′ formats discussed in the "BT.601 Video Interface" section.

Control Signals

In addition to the video data, there are four control signals:

HREF	horizontal blanking
VREF	vertical sync
VACTIVE	active video
PIXCLK	2 × sample clock

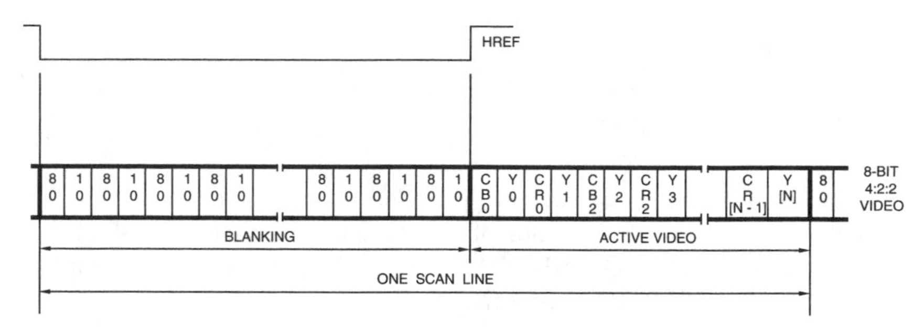

FIGURE 4.33 VMI 8-bit 4:2:2 YCbCr Data for One Scan Line.

BT.656 Interface

The BT.656 interface for ICs is based on the pro-video BT.656-type parallel interfaces, discussed earlier in this chapter. Using EAV and SAV sequences to indicate video timing reduces the number of pins required. The timing of the H, V, and F signals for common video formats is illustrated in Chapter 3.

Standard IC signal levels and timing are used, and any resolution can be supported.

Zoomed Video Port (ZV Port)

An early standard for notebook PCs, the ZV Port was a point-to-point unidirectional bus between the PC Card host adaptor and the graphics controller. It enabled video data to be transferred real time directly from the PC card into the graphics frame buffer.

Video Interface Port (VIP)

The VESA VIP specification is an enhancement to the BT.656 interface for ICs, previously discussed. The primary application is to interface up to four devices to a graphics controller chip, although the concept can easily be applied to other applications.

There are three sections to the interface:

Host Interface:

VIPCLK	host clock
HAD0–HAD7	host address/data bus
HCTL	host control

Video Interface:

PIXCLK	video sample clock
VID0–VID7	lower video data bus
VIDA, VIDB	10-bit data extension
XPIXCLK	video sample clock
XVID0–XVID7	upper video data bus
XVIDA, XVIDB	10-bit data extension

System Interface:

VRST#	reset
VIRQ#	interrupt request

The host interface signals are provided by the graphics controller. Essentially, a 2-, 4-, or 8-bit version of the PCI interface is used. VIP-CLK has a frequency range of 25–33 MHz. PIX-CLK and XPIXCLK have a maximum frequency of 75 and 80 MHz, respectively.

Consumer Component Interfaces

Many solutions for transferring digital video between equipment have been developed over the years. HDMI, originally derived from DVI, is the most popular digital video interface for consumer equipment.

Digital Visual Interface (DVI)

In 1998, the Digital Display Working Group (DDWG) was formed to address the need for a standardized digital video interface between a PC and VGA monitor, as illustrated in Figure 4.34. The DVI 1.0 specification was released in April 1999. Designed to transfer uncompressed real-time digital video, DVI supports PC graphics resolutions beyond 1600 × 1200 and HDTV resolutions, including 720p, 1080i, and 1080p.

In 2003, the consumer electronics industry started adding DVI outputs to DVD players and cable/satellite set-top boxes. DVI inputs also started appearing on digital televisions and LCD/plasma monitors.

Technology

DVI is based on the Digital Flat Panel (DFP) Interface, enhancing it by supporting more formats and timings. It also includes support for the High-bandwidth Digital Content Protection (HDCP) specification to deter unauthorized copying of content.

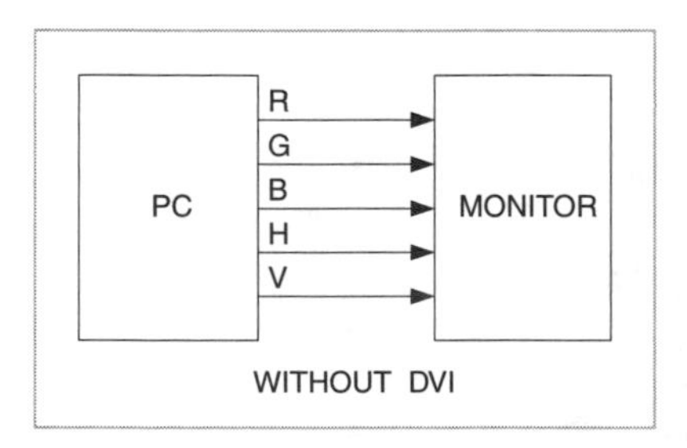

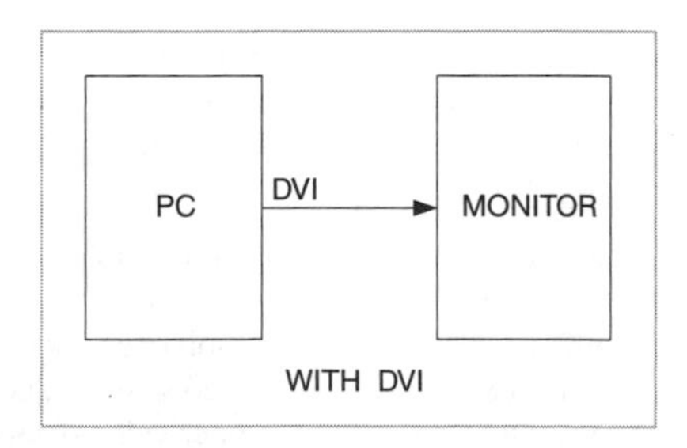

FIGURE 4.34 Using DVI to Connect a VGA Monitor to a PC.

DVI also supports VESA's Extended Display Identification Data (EDID) standard, Display Data Channel (DDC) standard (used to read the EDID), and Monitor Timing Specification (DMT).

DDC and EDID enable automatic display detection and configuration. Extended Display Identification Data (EDID) was created to enable plug and play capabilities of displays. Data is stored in the display, describing the supported video formats. This information is supplied to the source device, over DVI, at the request of the source device. The source device then chooses its output format, taking into account the format of the original video stream and the formats supported by the display. The source device is responsible for the format conversions necessary to supply video in an understandable form to the display.

In addition, the CEA-861 standard specifies mandatory and optionally supported resolutions and timings, and how to include data such as aspect ratio and format information.

How It Works

TMDS Links*: DVI uses* transition-minimized differential signaling *(TMDS). Eight bits of video data are converted to a 10-bit transition-minimized, DC-balanced value, which is then serialized. The receiver deserializes the data, and converts it back to 8 bits. Thus, to transfer digital R′G′B′ data requires three TMDS signals that comprise one TMDS link.*

"TFT data mapping" is supported as the minimum requirement: 1 pixel per clock, 8 bits per channel, MSB justified.

Either one or two TMDS links may be used, as shown in Figures 4.35 and 4.26, depending on the formats and timing required. A system supporting two TMDS links must be able to switch dynamically between formats requiring a single link and formats requiring

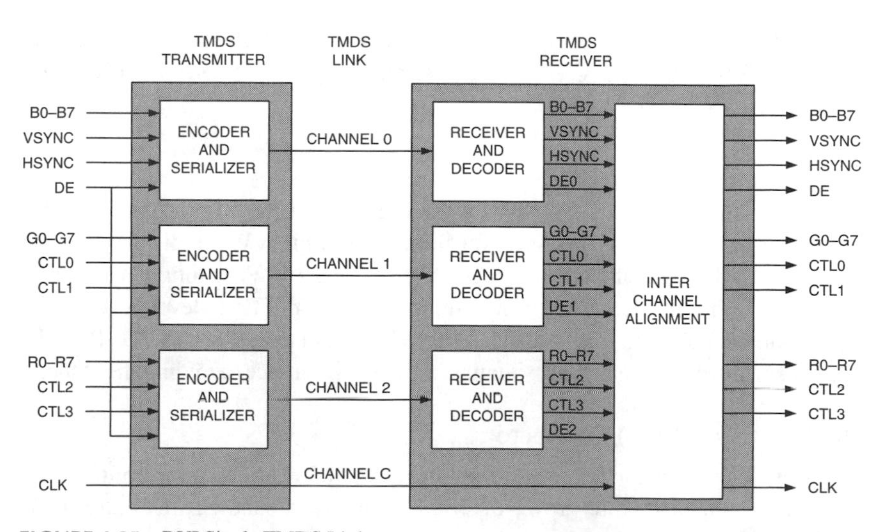

FIGURE 4.35 DVI Single TMDS Link.

a dual link. A single DVI connector can handle two TMDS links. A single TMDS link supports resolutions and timings using a video sample rate of 25–165 MHz. Resolutions and timings using a video sample rate of 165–330 MHz are implemented using two TMDS links, with each TMDS link operating at one-half the frequency. Thus, the two TMDS links share the same clock and the bandwidth is shared evenly between the two links.

Video Data Formats

Typically, 24-bit R′G′B′ data is transferred over a link. For applications requiring more than 8 bits per color component, the second TMDS link may be used for the additional least significant bits.

For PC applications, R′G′B′ data typically has a range of 0 × 00–0xFF. For consumer applications, R′G′B′ data typically has a range of 0 × 10–0xEB (values less than 0 × 10 or greater than 0xEB may be occasionally present due to processing).

Control Signals

In addition to the video data, DVI transmitter and receiver chips typically use up to 14 control signals for interfacing to other chips in the system:

HSYNC	horizontal sync
VSYNCVSYNC	vertical sync
DE	data enable
CTL0–CTL3	reserved (link 0)
CTL4–CTL9	reserved (link 1)
CLK	1 × sample clock

While DE is a "1," active video is processed. While DE is a "0," the HSYNC, VSYNC, and CTL0–CTL9 signals are processed. HSYNC and VSYNC may be either polarity.

Technology Trade-Offs

One issue is that some HDTVs use the falling edge of the YPbPr tri-level sync, rather than the center (rising edge), for horizontal timing. When displaying content from DVI, this results in the image shifting by 2.3%. Providing the ability to adjust the DVI embedded sync timing relative to the YPbPr tri-level sync timing is a useful capability in this case. Many fixed-pixel displays, such as DLP, LCD, and plasma, instead use the DE signal as a timing reference, avoiding the issue.

Digital-Only (DVI-D) Connector

The digital-only connector, which supports dual link operation, contains 24 contacts arranged as three rows of eight contacts, as shown in Figure 4.37. Table 4.30 lists the pin assignments.

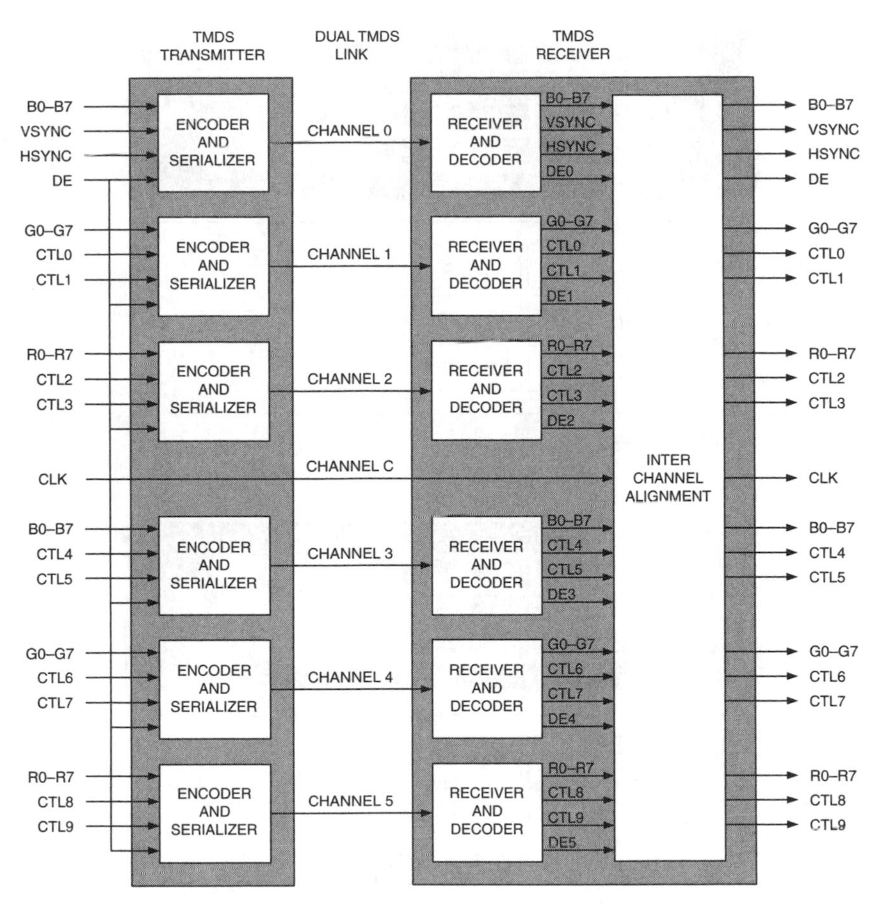

FIGURE 4.36 DVI Dual TMDS Link.

Digital-Analog (DVI-I) Connector

In addition to the 24 contacts used by the digital-only connector, the 29-contact digital-analog connector adds five additional contacts to support analog video as shown in Figure 4.38. Table 4.31 lists the pin assignments.

HSYNC	horizontal sync
VSYNC	vertical sync
RED	analog red video
GREEN	analog green video
BLUE	analog blue video

The operation of the analog signals is the same as for a standard VGA connector.

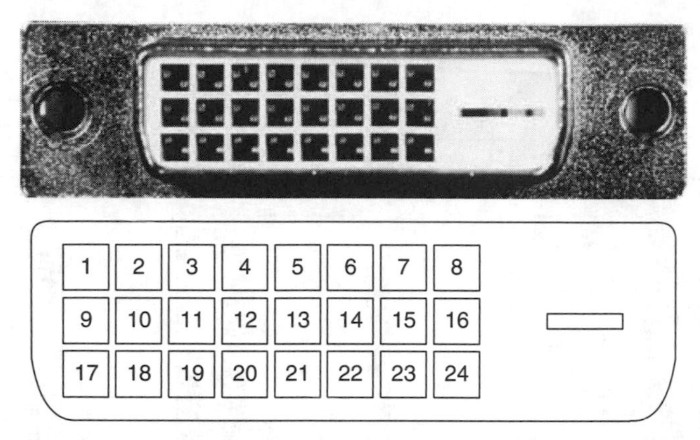

FIGURE 4.37 DVI-D Connector.

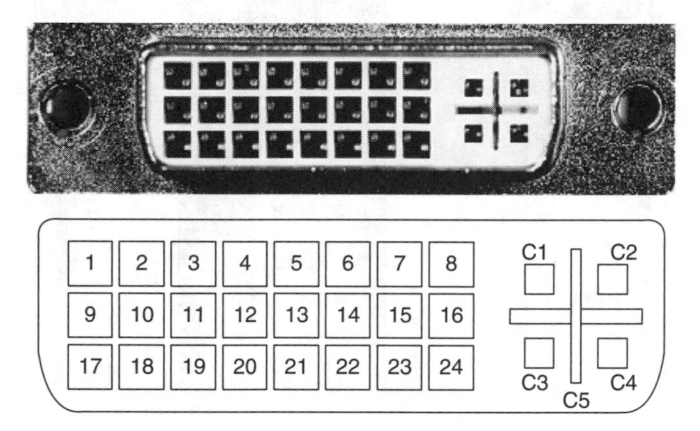

FIGURE 4.38 DVI-I Connector.

DVI-A is available as a plug (male) connector only and mates to the analog-only pins of a DVI-I connector. DVI-A is only used in adapter cables, where there is the need to convert to or from a traditional analog VGA signal.

High-Definition Multimedia Interface (HDMI)

Although DVI handles transferring uncompressed real-time digital RGB video to a display, the consumer electronics industry preferred a smaller, more flexible solution, based on DVI technology. In April 2002, the HDMI working group was formed by Hitachi, Matsushita Electric (Panasonic), Philips, Silicon Image, Sony, Thomson, and Toshiba.

HDMI is capable of replacing up to eight audio cables (7.1 channels) and up to three video cables with a single cable, as illustrated in Figure 4.39. In 2004, the consumer electronics industry started adding HDMI outputs to DVD

TABLE 4.31 **DVI-I Connector Signal Assignments**

Pin	Signal	Pin	Signal	Pin	Signal
1	D2−	**9**	D1−	**17**	D0−
2	D2	**10**	D1	**18**	D0
3	shield	**11**	shield	**19**	shield
4	D4−	**12**	D3−	**20**	D5−
5	D4	**13**	D3	**21**	D5
6	DDC SCL	**14**	+5V	**22**	shield
7	DDC SDA	**15**	ground	**23**	CLK
8	VSYNC	**16**	Hot Plug Detect	**24**	CLK−
C1	RED	**C2**	GREEN	**C3**	BLUE
C4	HSYNC	**C5**	ground		

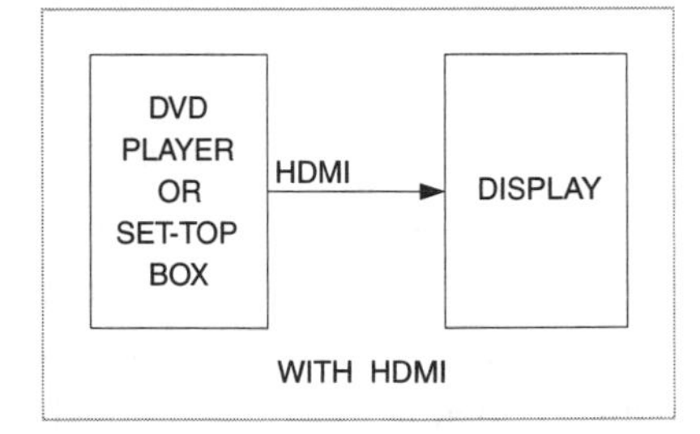

FIGURE 4.39 Using HDMI Eliminates Confusing Cable Connections for Consumers.

players and cable/satellite set-top boxes. HDMI inputs started appearing on digital televisions and monitors in 2005.

Insider Info

Through the use of an adaptor cable, HDMI is backwards compatible with equipment using DVI and the CEA-861 DTV profile. However, the advanced features of HDMI, such as digital audio, Consumer Electronics Control (used to enable passing control commands between equipment) and color gamut metadata, are not available.

Technology

HDMI, based on DVI, supports VESA's Extended Display Identification Data (EDID) standard and Display Data Channel (DDC) standard (used to read the EDID).

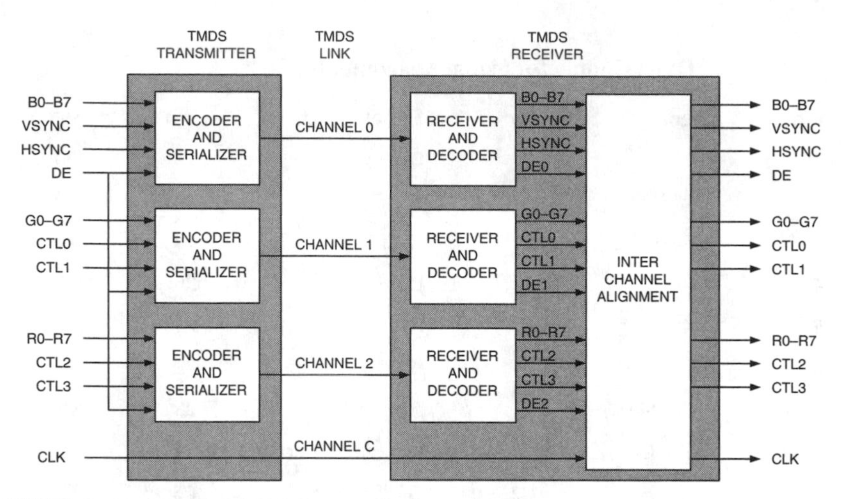

FIGURE 4.40 DFP TMDS Link.

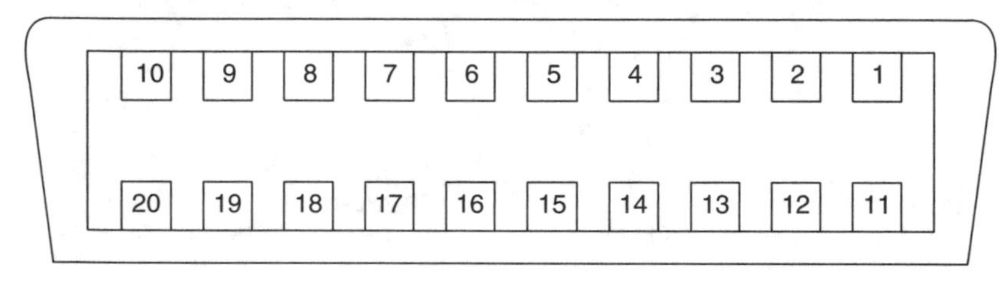

FIGURE 4.41 DFP Connector.

In addition, the CEA-861 standard specifies mandatory and optionally supported resolutions and timings, and how to include data such as aspect ratio and format information. HDMI also supports the High-bandwidth Digital Content Protection (HDCP) specification to deter unauthorized copying of content. A common problem is sources not polling the TV often enough (twice per second) to see if its HDCP circuit is active. This results in snow if the TV's HDMI input is deselected, then later selected again.

The 19-pin Type A connector uses a single TMDS link and can therefore carry video signals with a 25–340 MHz sample rate. Video with sample rates below 25 MHz (i.e. 13.5 MHz 480i and 576i) are transmitted using a pixel-repetition scheme.

To support video signals sampled at greater than 340 MHz, the dual-link capability of the 29-pin Type B connector is used.

The 19-pin Type C connector, designed for mobile applications, is a smaller version of the Type A connector.

Connector

The 20-pin mini-D ribbon (MDR) connector contains 20 contacts arranged as two rows of ten contacts, as shown in Table 4.32.

TABLE 4.32 **DFP Connector Signal Assignments**

Pin	Signal	Pin	Signal
1	D1	11	D2
2	D1–	12	D2–
3	shield	13	shield
4	shield	14	shield
5	CLK	15	D0
6	CLK–	16	D0–
7	ground	17	no ground
8	+5V	18	Hot Plug Detect
9	no connect	19	DDC SDA
10	no connect	20	DDC SCL

Open LVDS Display Interface (OpenLDI)

OpenLDI was developed for transferring uncompressed digital video from a computer to a digital flat panel display. It enhances the FPD-Link standard used to drive the displays of laptop computers, and adds support for VESA's Plug and Display (P&D) standard, Extended Display Identification Data (EDID) standard, and Display Data Channel (DDC) standard. DDC and EDID enable automatic display detection and configuration.

Unlike DVI and DFP, OpenLDI uses low-voltage differential signaling (LVDS). Cable lengths may be up to 10 meters.

LVDS Link

The LVDS link, as shown in Figure 4.42, supports formats and timings requiring a clock rate of 32.5–160 MHz.

Eight serial data lines (A0–A7) and two sample clock lines (CLK1 and CLK2) are used. The number of serial data lines actually used is dependent on the pixel format, with the serial data rate being 7x the sample clock rate. The CLK2 signal is used in the dual pixel modes for backwards compatibility with FPD-Link receivers.

Gigabit Video Interface (GVIF)

The Sony GVIF was developed for transferring uncompressed digital video using a single differential signal, instead of the multiple signals that DVI, DFP, and OpenLDI use. Cable lengths may be up to 10 meters.

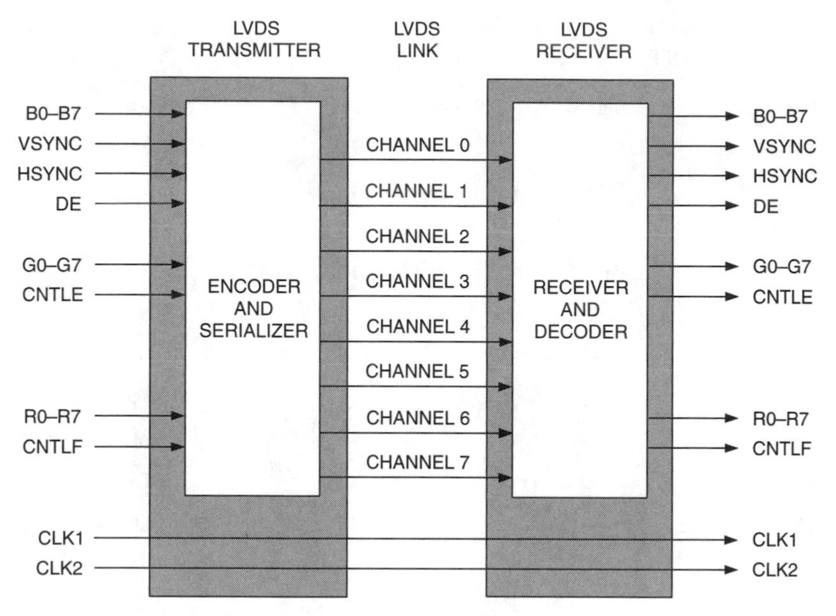

FIGURE 4.42 OpenLDI LVDS Link.

GVIF Link

The GVIF link, as shown in Figure 4.43, supports formats and timings requiring a clock rate of 20–80 MHz. For applications requiring higher clock rates, more than one GVIF link may be used.

The serial data rate is 24 × the sample clock rate for 18-bit R′G′B′ data, or 30 × the sample clock rate for 24-bit R′G′B′ data.

Video Data Formats

18-bit or 24-bit R′G′B′ data, plus timing, is transferred over the link. The 18-bit R′G′B′ format uses three 6-bit R′G′B′ values: R0–R5, G0–G5, and B0–B5. The 24-bit R′G′B′ format uses three 8-bit R′G′B′ values: R0–R7, G0–G7, and B0–B7.

18-bit R′G′B′ data is converted to 24-bit data by slicing the R´G´B data into six 3-bit values that are in turn transformed into six 4-bit codes. This ensures rich transitions for receiver PLL locking and good DC balance.

24-bit R′G′B′ data is converted to 30-bit data by slicing the R´G´B data into six 4-bit values that are in turn transformed into six 5-bit codes.

Control Signals

In addition to the video data, there are six control signals:

HSYNC	horizontal sync
VSYNC	vertical sync

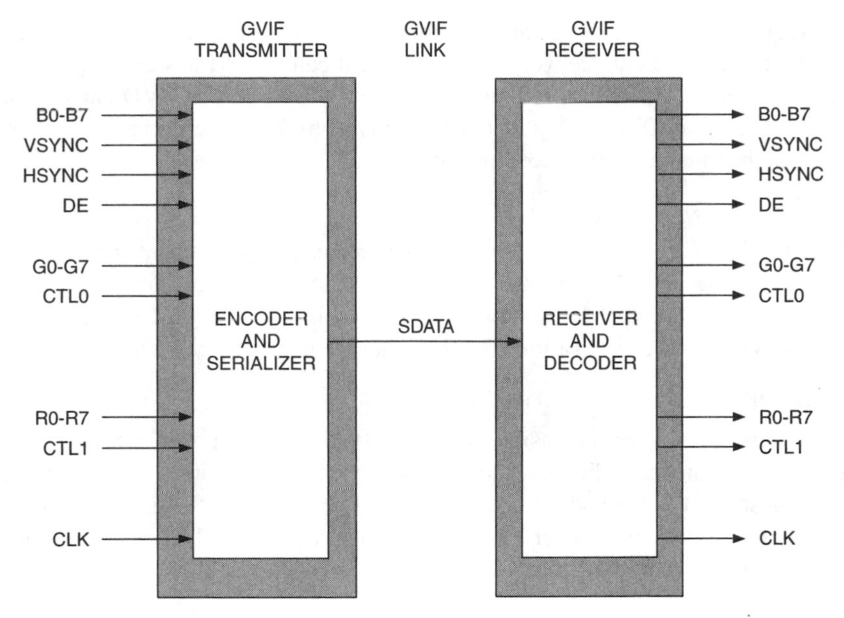

FIGURE 4.43 GVIF Link.

DE	data enable
CTL0	reserved
CTL1	reserved
CLK	1 × sample clock

If any of the HSYNC, VSYNC, DE, CTL0, or CTL1 signals change, during the next CLK cycle a special 30-bit format is used. The first 6 bits are header data indicating the new levels of HSYNC, VSYNC, DE, CTL0, or CTL1. This is followed by 24 bits of R′G′B′ data (unencoded except for inverting the odd bits).

Note that during the blanking periods, nonvideo data, such as digital audio, may be transferred. The CTL signals may be used to indicate when non-video data is present.

Consumer Transport Interfaces

Several transport interfaces, such as USB 2.0, Ethernet, and IEEE 1394, are available for consumer products. Of course, each standard has its own advantages and disadvantages.

USB 2.0

Well known in the PC market for connecting peripherals to a PC, there is growing interest in using USB (Universal Serial Bus) 2.0 to transfer compressed audio/video data between products.

USB 2.0 is capable of operating up to 480 Mbps and supports an isochronous mode to guarantee data delivery timing. Thus, it can easily transfer compressed real-time audio/ video data from a cable/satellite set-top box or DVD player to a digital television. DTCP (Digital Transmission Copy Protection) may be used to encrypt the audio and video content over USB.

Technology Trade-Offs

Due to USB's lower cost and widespread usage, many companies are interested in using USB 2.0 instead of IEEE 1394 to transfer compressed audio/ video data between products. However, some still prefer IEEE 1394 since the methods for transferring various types of data are much better defined.

USB On-the-Go

With portable devices increasing in popularity, there was a growing desire for them to communicate directly with each other without requiring a PC or other USB host. On-the-Go addresses this desire by allowing a USB device to communicate directly with other On-the-Go products. It also features a smaller USB connector and low power features to preserve battery life.

Ethernet

With the widespread adoption of home networks, DSL, and FTTH (Fiber-to-the-Home), Ethernet has become a common interface for transporting digital audio and video data. Initially used for file transfers, streaming of real-time compressed video over wired (802.3) or wireless (802.11) Ethernet networks is now becoming common.

Ethernet supports up to 1 Gbps. DTCP/IP (Digital Transmission Copy Protection for Internet Protocol) may be used to encrypt the audio and video content over wired or wireless networks.

IEEE 1394

IEEE 1394 was originally developed by Apple Computer as Firewire. Designed to be a generic interface between devices, 1394 specifies the physical characteristics; separate application-specific specifications describe how to transfer data over the 1394 network.

1394 is a transaction-based packet technology, using a bi-directional serial interconnect that features hot plug-and-play. This enables devices to be connected and disconnected without affecting the operation of other devices connected to the network.

Guaranteed delivery of time-sensitive data is supported, enabling digital audio and video to be transferred in real time. In addition, multiple independent streams of digital audio and video can be carried.

Specifications

The original 1394-1995 specification supports bit-rates of 98.304, 196.608, and 393.216 Mbps.

The 1394A-2000 specification clarifies areas that were vague and led to system interoperability issues. It also reduces the overhead lost to bus control, arbitration, bus reset duration, and concatenation of packets. 1394A-2000 also introduces advanced power-saving features. The electrical signaling method is also common between 1394–1995 and 1394A-2000, using data-strobe (DS) encoding and analog-speed signaling.

The 1394B-2002 specification adds support for bit-rates of 786.432, 1572.864, and 3145.728 Mbps. It also includes

- 8B/10B encoding technique used by Gigabit Ethernet
- Continuous dual simplex operation
- Longer distance (up to 100 meters over Cat5)
- Changes the speed signaling to a more digital method
- Three types of ports: Legacy (1395A compatible), Beta, and Bilingual (supports both Legacy and Beta). Connector keying ensures that incompatible conncctions cannot physically be made.

Endian Issues: 1394 uses a big-endian architecture, defining the most significant bit as bit 0. However, many processors are based on the little-endian architecture which defines the most significant bit as bit 31 (assuming a 32-bit word).

Network Topology

Like many networks, there is no designated bus master. The tree-like network structure has a root node, branching out to logical nodes in other devices (Figure 4.44). The root is responsible for certain control functions, and is

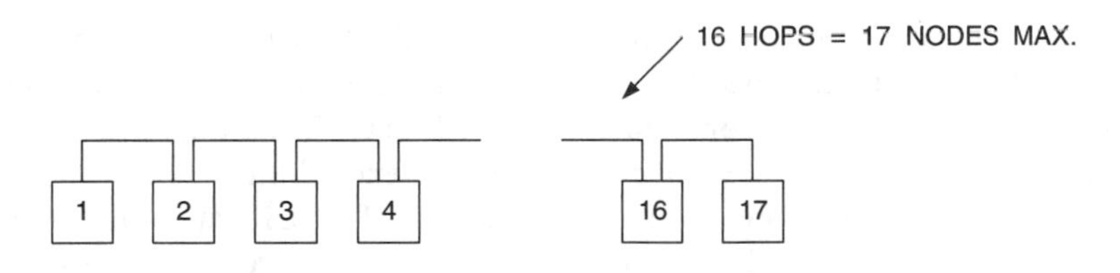

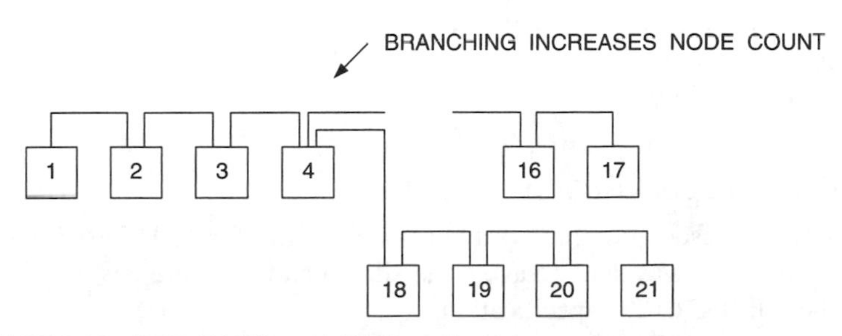

FIGURE 4.44 IEEE 1394 Network Topology Examples.

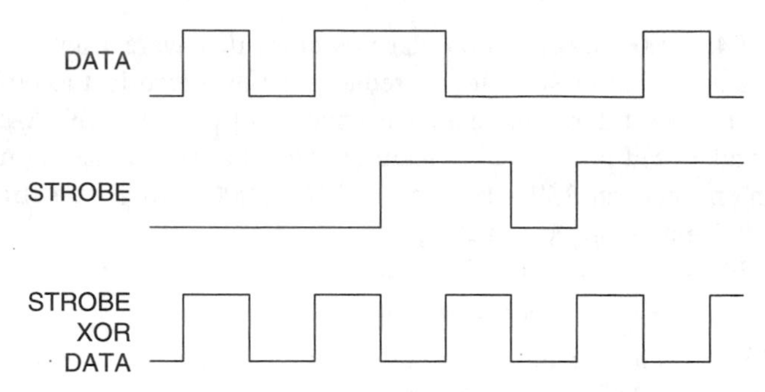

FIGURE 4.45 IEEE 1394 Data and Strobe Signal Timing.

chosen during initialization. Once chosen, it retains that function for as long as it remains powered on and connected to the network.

A network can include up to 63 nodes, with each node (or device) specified by a 6-bit physical identification number. Multiple networks may be connected by bridges, up to a system maximum of 1,023 networks, with each network represented by a separate 10-bit bus ID. Combined, the 16-bit address allows up to 64,449 nodes in a system. Since device addresses are 64 bits, and 16 of these bits are used to specify nodes and networks, 48 bits remain for memory addresses, allowing up to 256TB of memory space per node.

Digital Transmission Content Protection (DTCP)

To prevent unauthorized copying of content, the DTCP system was developed. Although originally designed for 1394, it is applicable to any digital network that supports bidirectional communications, such as USB and Ethernet.

Device authentication, content encryption, and renewability (should a device ever be compromised) are supported by DTCP. The Digital Transmission Licensing Administrator (DTLA) licenses the content protection system and distributes cipher keys and device certificates.

DTCP outlines four elements of content protection:

1. Copy control information (CCI)
2. Authentication and key exchange (AKE)
3. Content encryption
4. System renewability

Digital Camera Specification

The 1394 Trade Association has written a specification for 1394-based digital video cameras. This was done to avoid the silicon and software cost of implementing the full IEC 61883 specification.

Seven resolutions are defined, with a wide range of format support:

160 × 120	4:4:4 YCbCr
320 × 240	4:2:2 YCbCr
640 × 480	4:1:1, 4:2:2 YCbCr, 24-bit RGB
800 × 600	4:2:2 YCbCr, 24-bit RGB
1024 × 768	4:2:2 YCbCr, 24-bit RGB
1280 × 960	4:2:2 YCbCr, 24-bit RGB
1600 × 1200	4:2:2 YCbCr, 24-bit RGB

Supported frame rates are 1.875, 3.75, 7.5, 15, 30, and 60 frames per second.

Isochronous packets are used to transfer the uncompressed digital video data over the 1394 network.

INSTANT SUMMARY

Video interfaces make it possible to exchange video information between devices. Interface standards include:

Analog video interfaces

- S-Video
- SCART
- SDTV RGB interface
- HDTV RGB interface
- SDTV YPbPr interface
- HDTV YPbPr interface
- D-Connector interface
- VGA interface

Digital video interfaces

- Pro-Video component interfaces
- Pro-Video composite interfaces
- Pro-Video Transport interfaces
- IC component interfaces
- Consumer component interfaces
- Consumer Transport interfaces

Chapter 5

Digital Video Processing

In an Instant

- Processing definitions
- Display enhancement
- Video mixing and graphics overlay
- Chroma keying
- Video scaling
- Scan rate conversion
- Noninterlaced-to-interlaced conversion
- Interlaced-to-noninterlaced conversion
- DCT-based compression

Processing Definitions

Besides *encoding and decoding* MPEG, NTSC/PAL, and many other types of video, a typical system usually requires considerable additional video processing.

For example, since many consumer displays, and most computer displays, are progressive (noninterlaced), interlaced video must be converted to progressive (*interlaced-to-noninterlaced*). Progressive video must be converted to interlaced to drive a conventional analog VCR or interlaced TV, requiring *noninterlaced-to-interlaced* conversion.

Many computer displays support refresh rates up to at least 75 frames per second. CRT-based televisions have a refresh rate of 50 or 59.94 (60/1.001) fields per second, and refresh rates of up to 120 frames per second are becoming common for flat-panel televisions. For film-based compressed content, the source may only be 24 frames per second. Thus, some form of *frame rate conversion*, or *scan rate conversion*, must be done.

Another not-so-subtle problem includes *video scaling*. SDTV and HDTV support multiple resolutions, yet the display may be a single, fixed resolution.

Alpha mixing and *chroma keying* are used to mix multiple video signals or video with computer-generated text and graphics. Alpha mixing ensures a smooth crossover between sources, allows subpixel positioning of text, and limits source transition bandwidths to simplify eventual encoding to composite video signals.

Since no source is perfect, even digital sources, user controls for adjustable *brightness*, *contrast*, *saturation*, and *hue* are always desirable.

DISPLAY ENHANCEMENT

Brightness, Contrast, Saturation (Color), and Hue (Tint)

Working in the YCbCr color space simplifies the implementation of brightness, contrast, saturation, and hue controls, as shown in Figure 5.1. Also illustrated are multiplexers to allow the output of black screen, blue screen, and color bars.

ALERT!

The designer should ensure that no overflow or underflow wraparound errors occur, effectively saturating results to the 0 and 255 values.

Many displays also use separate hue and saturation controls for each of the red, green, blue, cyan, yellow, and magenta colors. This enables tuning the image at production time to better match the display's characteristics.

Color Transient Improvement

YCbCr transitions should be aligned. However, the Cb and Cr transitions are usually slower and time-offset due to the narrower bandwidth of color difference information.

By monitoring coincident Y transitions, faster horizontal and vertical transitions may be synthesized for Cb and Cr. Small pre- and after-shoots may also be added to the Cb and Cr signals.

The new Cb and Cr edges are then aligned with the Y edge, as shown in Figure 5.2.

Displays commonly use this technique to provide a sharper-looking picture.

Luma Transient Improvement

In this case, the Y horizontal and vertical transitions are shortened, and small pre- and after-shoots may also be added, to artificially sharpen the image.

Displays commonly use this technique to provide a sharper-looking picture.

Sharpness

The apparent sharpness of a picture may be increased by increasing the amplitude of high-frequency luminance information.

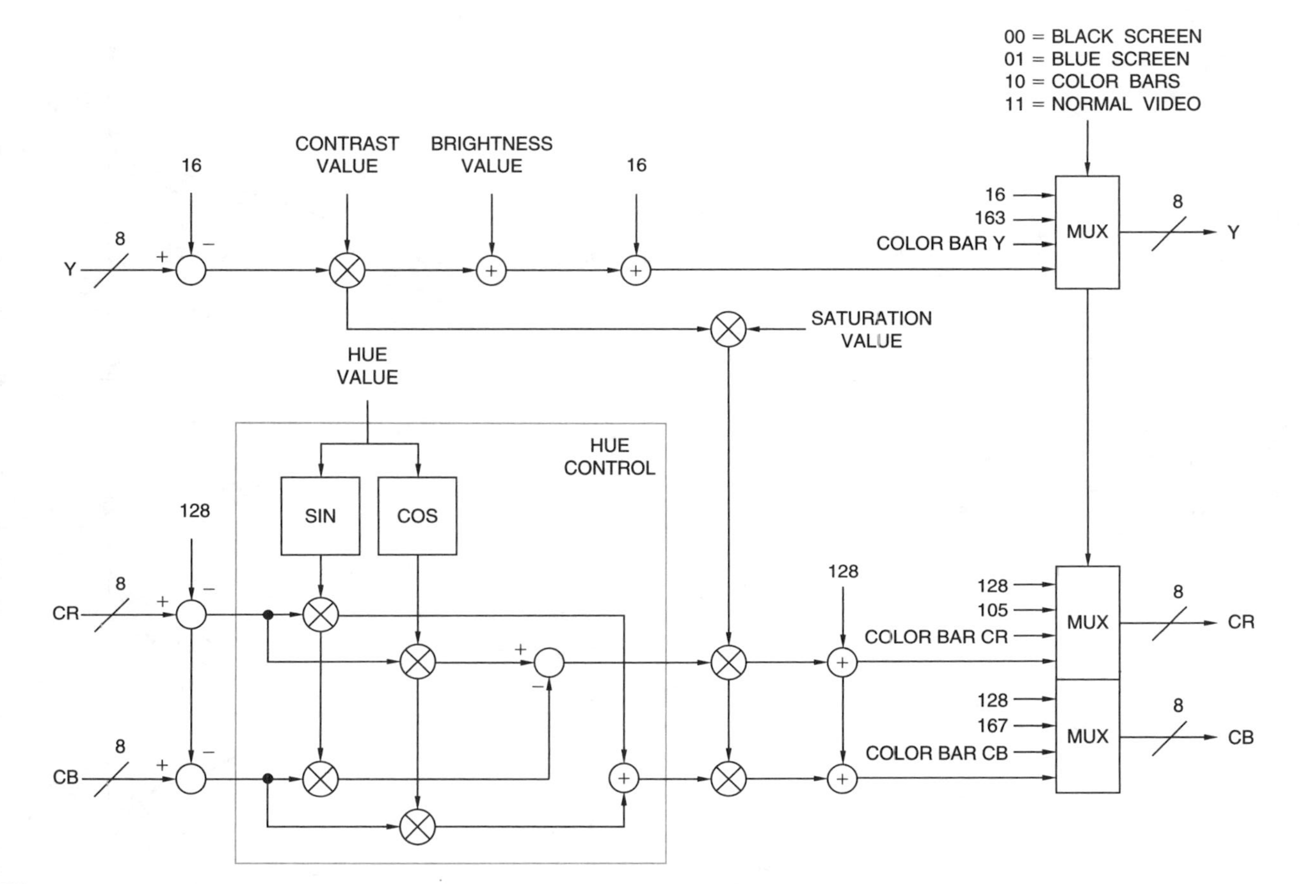

FIGURE 5.1 Hue, Saturation, Contrast, and Brightness Controls.

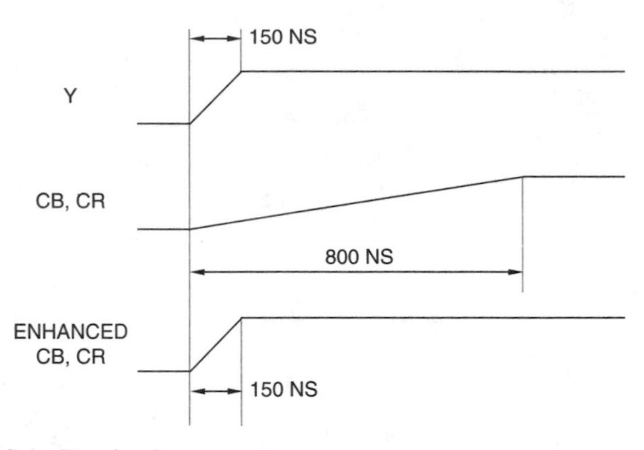

FIGURE 5.2 Color Transient Improvement.

How It Works

As shown in Figure 5.3, a simple bandpass filter with selectable gain (also called a peaking filter) may be used. The frequency where maximum gain occurs is usually selectable to be either at the color subcarrier frequency or at about 2.6 MHz. A coring circuit is typically used after the filter to reduce low-level noise.

Figure 5.4 illustrates a more complex sharpness control circuit. The high-frequency luminance is increased using a variable band-pass filter, with adjustable gain. The coring function (typically ±1 LSB) removes low-level noise. The modified luminance is then added to the original luminance signal.

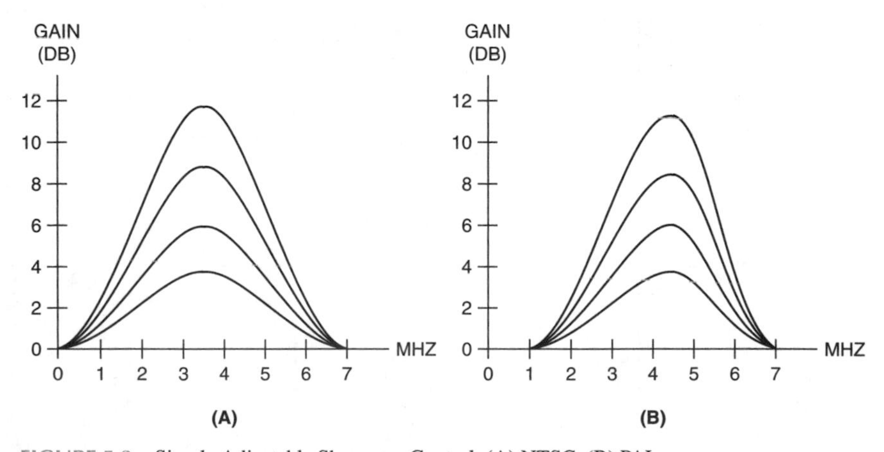

FIGURE 5.3 Simple Adjustable Sharpness Control. (A) NTSC. (B) PAL.

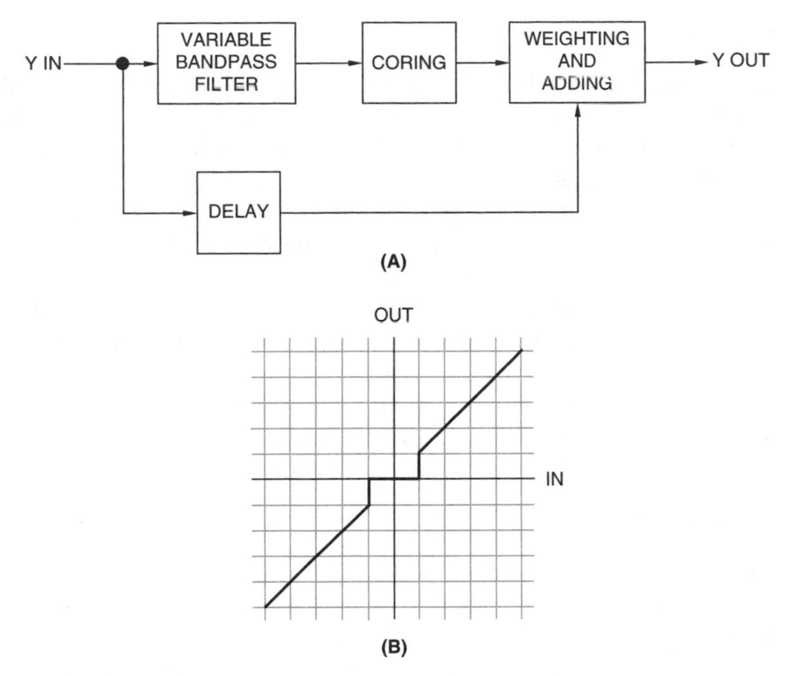

FIGURE 5.4 More Complex Sharpness Control. (A) Typical implementation. (B) Coring function.

Blue Stretch

Blue stretch increases the blue value of white and near-white colors in order to make whites appear brighter. When applying blue stretch, only colors within a specified color range should be processed.

Colors with a Y value of ~80% or more of the maximum with a low saturation value and that fall within a white detection area in the CbCr-plane, have their blue components increased by ~4% (the blue gain factor) and their red components decreased the same amount. For more complex designs, the white detection area and blue gain factor can be dependent on the color's Y value and saturation level.

Green Enhancement

Green enhancement creates a richer, more saturated green color when the level of green is low. Displays commonly use this technique to provide greener looking grass, plants, etc. When applying green enhancement, only colors only within a specified color range should be processed.

Dynamic Contrast

Using *dynamic contrast* (also called *adaptive contrast enhancement*), the differences between dark and light portions of the image are artificially enhanced based on the content in the image. Displays commonly use this technique to improve their contrast ratio.

Bright colors in mostly dark images are enhanced by making them brighter (white stretch). This is typically done by using histogram information to modify the upper portion of the gamma curve.

Dark colors in mostly light images are enhanced by making them darker (black stretch). This is typically done by using histogram information to modify the lower portion of the gamma curve.

For a medium-bright image, both techniques may be applied.

A minor gamma correction adjustment may also be applied to colors that are between dark and light, resulting in a more detailed and contrasting picture.

Color Correction

The RGB chromaticities are usually slightly different between the source video and what the display uses. This results in red, green and blue colors that are not completely accurate.

Color correction can be done on the source video to compensate for the display characteristics, enabling more accurate red, green and blue colors to be displayed.

An alternate type of color correction is to perform color expansion, taking advantage of the greater color reproduction capabilities of modern displays. This can result in greener greens, bluer blues, etc. One common technique of implementing color expansion is to use independent hue and saturation controls for each primary and complementary color, plus the skin color.

VIDEO MIXING AND GRAPHICS OVERLAY

Mixing video signals may be as simple as switching between two video sources. This is adequate if the resulting video is to be displayed on a computer monitor.

For most other applications, a technique known as *alpha mixing* should be used. Alpha mixing may also be used to fade to or from a specific color (such as black) or to overlay computer-generated text and graphics onto a video signal.

ALERT!

Alpha mixing must be used if the video is to be encoded to composite video. Otherwise, ringing and blurring may appear at the source switching points, such as around the edges of computer-generated text and graphics. This is due to the color information being lowpass filtered within the NTSC/PAL encoder. If the filters have a sharp cut-off, a fast color transition will produce ringing. In addition, the intensity information may be bandwidth-limited to about 4–5 MHz somewhere along the video path, slowing down intensity transitions.

Mathematically, with alpha normalized to have values of 0–1, alpha mixing is implemented as:

$$Out = (alpha\_0)(in\_0) + (alpha\_1)(in\_1) + \ldots$$

In this instance, each video source has its own alpha information. The alpha information may not total to one (unity gain).

Technology Trade-offs

A common problem in computer graphics systems that use alpha is that the frame buffer may contain preprocessed R′G′B′ or YCbCr data; that is, the R′G′B′ or YCbCr data in the frame buffer has already been multiplied by alpha. Assuming an alpha (A) value of 0.5, nonprocessed R′G′B′A values for white are (255, 255, 255, 128); preprocessed R′G′B′A values for white are (128, 128, 128, 128). Therefore, any mixing circuit that accepts R′G′B′ or YCbCr data from a frame buffer should be able to handle either format.

If using alpha mixing for special effects, such as wipes, the switching point (where 50% of each video source is used) must be able to be adjusted to an accuracy of less than one sample to ensure smooth movement. By controlling the alpha values, the switching point can be effectively positioned anywhere.

Text can be overlaid onto video by having a character generator control the alpha inputs. By setting one of the input sources to a constant color, the text will assume that color.

Insider Info

Note that for those designs that subtract 16 (the black level) from the Y channel before processing, negative Y values should be supported after the subtraction. This allows the design to pass through real-world and test video signals with minimum artifacts.

CHROMA KEYING

Chroma keying involves specifying a desired foreground key color; foreground areas containing the key color are replaced with the background image. Cb and Cr are used to specify the key color; luminance information may be used to increase the realism of the chroma keying function. The actual mixing of the two video sources may be done in the component or composite domain, although component mixing reduces artifacts.

Early chroma keying circuits simply performed a hard or soft switch between the fore-ground and background sources. In addition to limiting the amount of fine detail maintained in the foreground image, the background was not visible through transparent or translucent foreground objects, and shadows from the foreground were not present in areas containing the background image.

Linear keyers were developed that combine the foreground and background images in a proportion determined by the key level, resulting in the foreground image being attenuated in areas containing the background image. Although allowing foreground objects to appear transparent, there is a limit on the fineness of detail maintained in the foreground. Shadows from the foreground are not present in areas containing the background image unless additional processing is done—the luminance levels of specific areas of the background image must be reduced to create the effect of shadows cast by foreground objects.

If the blue or green backing used with the foreground scene is evenly lit except for shadows cast by the foreground objects, the effect on the background will be that of shadows cast by the foreground objects. This process, referred to as *shadow chroma keying*, or *luminance modulation*, enables the background luminance levels to be adjusted in proportion to the brightness of the blue or green backing in the foreground scene. This results in more realistic keying of transparent or translucent foreground objects by preserving the spectral highlights.

Note that green backgrounds are now more commonly used due to lower chroma noise.

Chroma keyers are also limited in their ability to handle foreground colors that are close to the key color without switching to the background image. Another problem may be a bluish tint to the foreground objects as a result of blue light reflecting off the blue backing or being diffused in the camera lens. Chroma spill is difficult to remove since the spill color is not the original key color; some mixing occurs, changing the original key color slightly.

Insider Info

One solution to many of the chroma keying problems is to process the foreground and background images individually before combining them, as shown in Figure 5.5. Rather than choosing between the foreground and background, each is processed individually and then combined. Figure 5.6 illustrates the major processing steps for both the foreground and background images during the chroma key process. Not shown in Figure 5.5 is the circuitry to initially subtract 16 (Y) or

VIDEO SCALING

With all the various video resolutions (Table 5.1), scaling is usually needed in almost every solution.

When generating objects that will be displayed on SDTV, computer users must be concerned with such things as text size, line thickness, and so forth. For example, text readable on a 1280 × 1024 computer display may not be readable on an SDTV display due to the large amount of downscaling involved. Thin horizontal lines may either disappear completely or flicker at a 25 or 29.97 Hz rate when converted to interlaced SDTV.

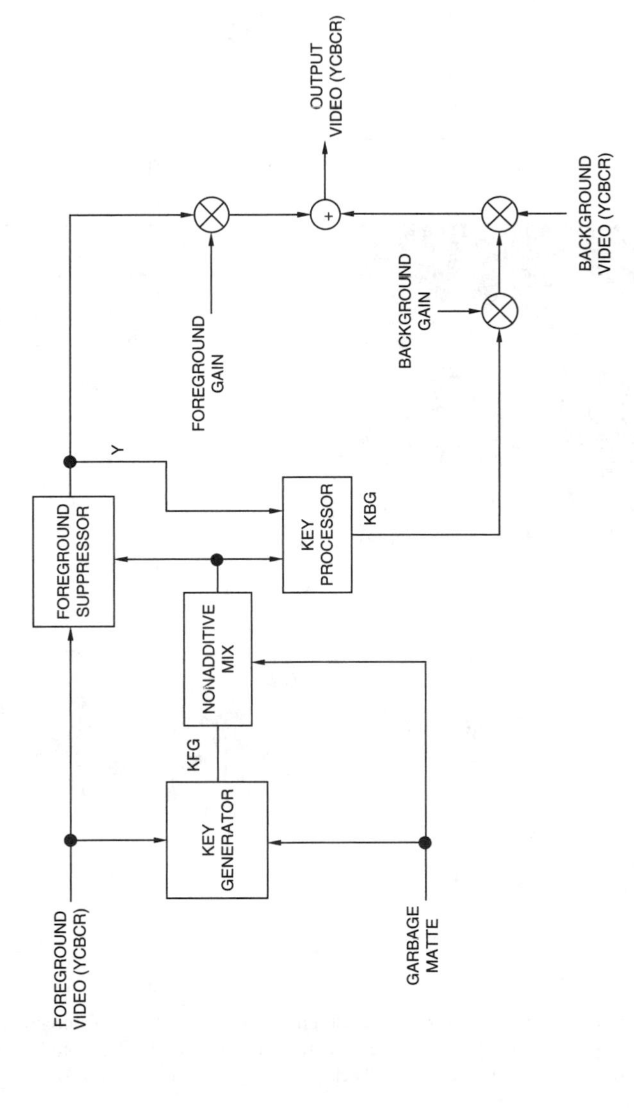

FIGURE 5.5 Typical Component Chroma Key Circuit.

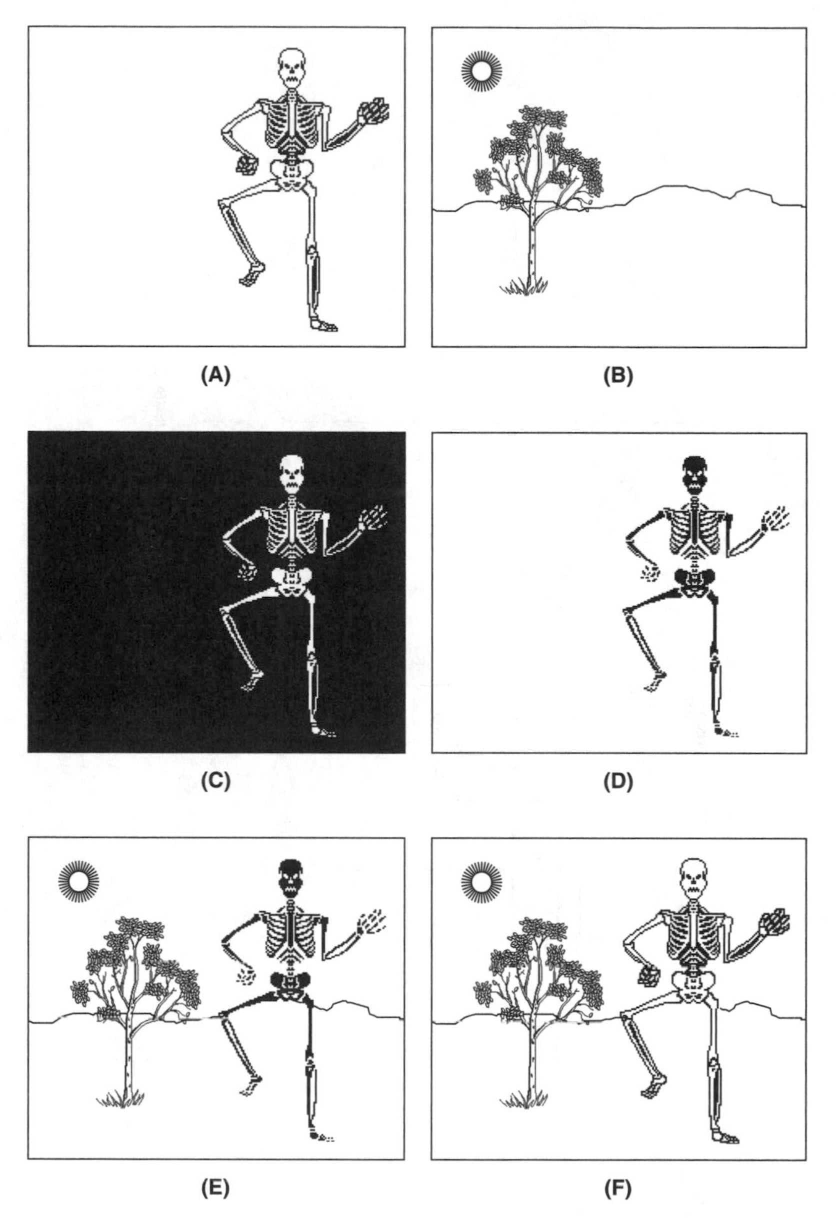

FIGURE 5.6 Major Processing Steps During Chroma Keying. (A) Original foreground scene. (B) Original background scene. (C) Suppressed foreground scene. (D) Background keying signal. (E) Background scene after multiplication by background key. (F) Composite scene generated by adding (C) and (E).

TABLE 5.1 Common Active Resolutions for Consumer Displays and Broadcast Sources. [1]16:9 letterbox on a 4:3 display. [2]2.35:1 anamorphic for a 16:9 1920 × 1080 display. [3]1.85:1 anamorphic for a 16:9 1920 × 1080 display

Displays		SDTV Sources		HDTV Sources
704 × 480	640 × 480	704 × 360[1]	704 × 432[1]	1280 × 720
854 × 480	800 × 600	480 × 480	480 × 576	1440 × 816[2]
704 × 576	1024 × 768	528 × 480		1440 × 1040[3]
854 × 576	1280 × 768	544 × 480	544 × 576	1280 × 1080
1280 × 720	1366 × 768	640 × 480		1440 × 1080
1280 × 768	1024 × 1024	704 × 480	704 × 576	1920 × 1080
1920 × 1080	1280 × 1024		768 × 576	

ALERT!

Note that scaling must be performed on component video signals (such as R′G′B′ or YCbCr). Composite color video signals cannot be scaled directly due to the color subcarrier phase information present, which would be meaningless after scaling.

In general, the spacing between output samples can be defined by a Target Increment (tarinc) value:

$$\text{tarinc} = I/O$$

where I and O are the number of input (I) and output (O) samples, either horizontally or vertically.

The first and last output samples may be aligned with the first and last input samples by adjusting the equation to be:

$$\text{tarinc} = (I - 1)/(O - 1)$$

Table 5.1 Common Active Resolutions for Consumer Displays and Broadcast Sources. 116:9 letterbox on a 4:3 display. 22.35:1 anamorphic for a 16:9 1920 × 1080 display. 31.85:1 anamorphic for a 16:9 1920 × 1080 display.

Pixel Dropping and Duplication

This is also called "nearest neighbor" scaling since only the input sample closest to the output sample is used.

The simplest form of scaling down is pixel dropping, where (m) out of every (n) samples are thrown away both horizontally and vertically. A modified version of the Bresenham line-drawing algorithm (described in most computer graphics books) is typically used to determine which samples not to discard.

Simple upscaling can be accomplished by pixel duplication, where (m) out of every (n) samples are duplicated both horizontally and vertically. Again, a modified version of the Bresenham line-drawing algorithm can be used to determine which samples to duplicate.

ALERT!

Scaling using pixel dropping or duplication is not recommended due to the visual artifacts and the introduction of aliasing components.

Linear Interpolation

An improvement in video quality of scaled images is possible using linear interpolation. When an output sample falls between two input samples (horizontally or vertically), the output sample is computed by linearly interpolating between the two input samples. However, scaling to images smaller than one-half of the original still results in deleted samples.

Figure 5.7 illustrates the vertical scaling of a 16:9 image to fit on a 4:3 display.

Technology Trade-offs

The linear interpolator is a poor bandwidth-limiting filter. Excess high-frequency detail is removed unnecessarily and too much energy above the Nyquist limit is still present, resulting in aliasing.

Anti-Aliased Resampling

The most desirable approach is to ensure the frequency content scales proportionally with the image size, both horizontally and vertically. Figure 5.8 illustrates the fundamentals of an anti-aliased resampling process. The input data is upsampled by A and lowpass filtered to remove image frequencies created by the interpolation process. Filter B bandwidth-limits the signal to remove frequencies that will alias in the resampling process B. The ratio of B/A determines the scaling factor.

Technology Trade-offs

Filters A and B are usually combined into a single filter. The response of the filter largely determines the quality of the interpolation. The ideal lowpass

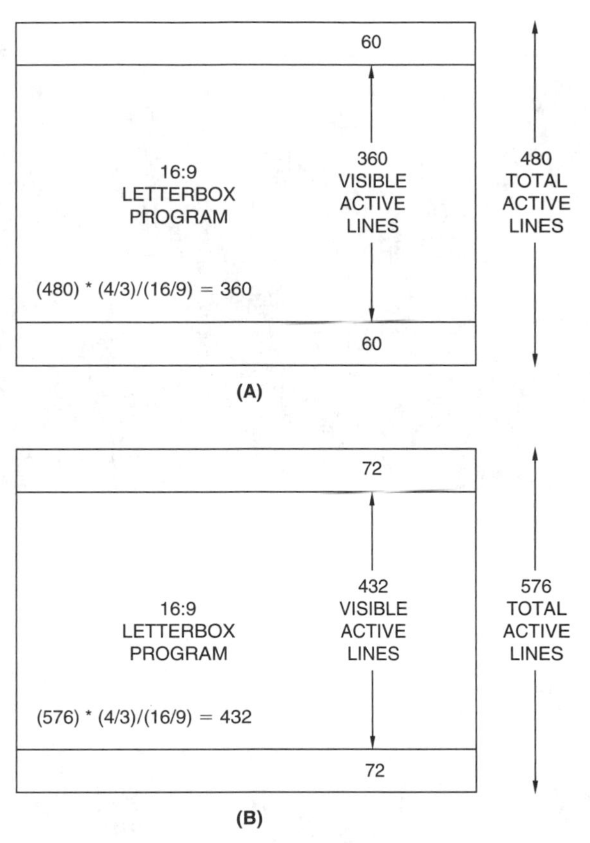

FIGURE 5.7 Vertical Scaling of 16:9 Images to Fit on a 4:3 Display. (A) 480-line systems. (B) 576-line systems.

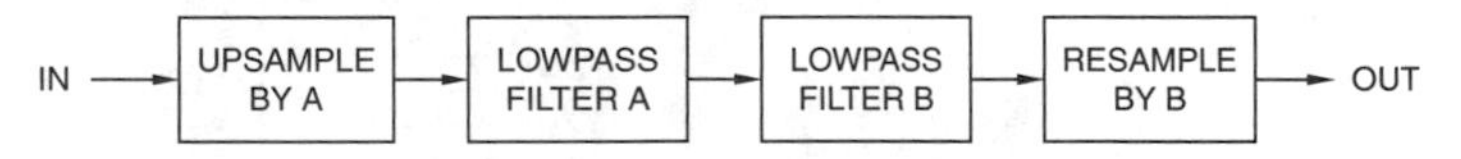

FIGURE 5.8 General Anti-Aliased Resampling Structure.

filter would have a very flat passband, a sharp cutoff at half of the lowest sampling frequency (either input or output), and very high attenuation in the stopband. However, since such a filter generates ringing on sharp edges, it is usually desirable to roll off the top of the passband. This makes for slightly softer pictures, but with less pronounced ringing.

Display Scaling Examples

Figures 5.9 through 5.17 illustrate various scaling examples for displaying 16:9 and 4:3 pictures on 4:3 and 16:9 displays, respectively.

FIGURE 5.9 16:9 Source Example.

FIGURE 5.10 Scaling 16:9 Content for a 4:3 Display: "Normal" or pan-and-scan mode. Results in some of the 16:9 content being ignored (indicated by gray regions).

How content is displayed is a combination of user preferences and content aspect ratio. For example, when displaying 16:9 content on a 4:3 display, many users prefer to have the entire display filled with the cropped picture (Figure 5.10) rather than seeing black or gray bars with the letterbox solution (Figure 5.11).

In addition, some displays incorrectly assume any progressive video signal on their YPbPr inputs is from an "anamorphic" source. As a result, they

FIGURE 5.11 Scaling 16:9 Content for a 4:3 Display: "Letterbox" mode. Entire 16:9 program visible, with black bars at top and bottom of display.

FIGURE 5.12 Scaling 16:9 Content for a 4:3 Display: "Squeezed" mode. Entire 16:9 program horizontally squeezed to fit 4:3 display, resulting in a distorted picture.

horizontally upscale progressive 16:9 programs by 25% when no scaling should be applied. Therefore, for set-top boxes it is useful to include a "16:9 (Compressed)" mode, which horizontally downscales the progressive 16:9 program by 25% to pre-compensate for the horizontally upscaling being done by the 16:9 display.

FIGURE 5.13 4:3 Source Example.

FIGURE 5.14 Scaling 4:3 Content for a 16:9 Display: "Normal" mode. Left and right portions of 16:9 display not used, so made black or gray.

SCAN RATE CONVERSION

In many cases, some form of *scan rate conversion* (also called temporal rate conversion, frame rate conversion, or field rate conversion) is needed. Multi-standard analog VCRs and scan converters use scan rate conversion to convert between various video standards. Computers usually operate the display at about 75 Hz noninterlaced, yet need to display 50 and 60 Hz interlaced video. With digital television, multiple frame rates can be supported.

FIGURE 5.15 Scaling 4:3 Content for a 16:9 Display: "Wide" mode. Entire picture linearly scaled horizontally to fill 16:9 display, resulting in distorted picture unless used with anamorphic content.

FIGURE 5.16 Scaling 4:3 Content for a 16:9 Display: "Zoom" mode. Top and bottom portion of 4:3 picture deleted, then scaled to fill 16:9 display.

ALERT!

Note that processing must be performed on component video signals (such as R′G′B′ or YCbCr). Composite color video signals cannot be processed directly due to the color subcarrier phase information present, which would be meaningless after processing.

FIGURE 5.17 Scaling 4:3 Content for a 16:9 Display: "Panorama" mode. Left and right 25% edges of picture are nonlinearly scaled horizontally to fill 16:9 display, distorted picture on left and right sides.

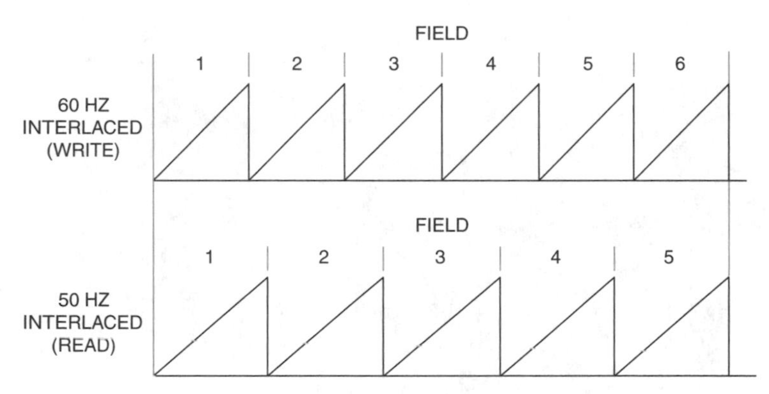

FIGURE 5.18 60 Hz to 50 Hz Conversion Using a Single Field Store by Dropping One out of Every Six Fields.

Frame or Field Dropping and Duplicating

Simple scan-rate conversion may be done by dropping or duplicating one out of every N fields. For example, the conversion of 60 Hz to 50 Hz interlaced operation may drop one out of every six fields, as shown in Figure 5.18, using a single field store.

Technology Trade-offs

The disadvantage of this technique is that the viewer may see jerky motion, or motion judder. In addition, some video decompression products use top-field only to convert from 60 Hz to 50 Hz, degrading the vertical resolution. The worst artifacts are present when a non-integer scan rate conversion is done—for example, when some frames are displayed three times, while others are displayed twice. In this instance, the viewer will observe double or blurred objects. As the human brain tracks an object in successive frames, it expects to see a regular sequence of positions, and has trouble reconciling the apparent stop-start motion of objects. As a result, it incorrectly concludes that there are two objects moving in parallel.

Temporal Interpolation

This technique generates new frames from the original frames as needed to generate the desired frame rate. Information from both past and future input frames should be used to optimally handle objects appearing and disappearing.

For every five fields of 50 Hz video, there will be six fields of 60 Hz video.

After both sources are aligned, two adjacent 50 Hz fields are mixed together to generate a new 60 Hz field. This technique is used in some inexpensive standards converters to convert between 50 Hz and 60 Hz standards. Note that no motion analysis is done. Therefore, if the camera operating at 50 Hz and 60 Hz pans horizontally past a narrow vertical object, you see one object once every six 60 Hz fields, and for the five fields in between, you see two objects, one fading in while the other fades out.

NONINTERLACED-TO-INTERLACED CONVERSION

In some applications, it is necessary to display a noninterlaced video signal on an interlaced display. Thus, some form of noninterlaced-to-interlaced conversion may be required.

Noninterlaced-to-interlaced conversion must be performed on component video signals (such as R′G′B′ or YCbCr). Composite color video signals (such as NTSC or PAL) cannot be processed directly due to the presence of color subcarrier phase information, which would be meaningless after processing. These signals must be decoded into component color signals, such as R′G′B′ or YCbCr, prior to conversion.

There are essentially two techniques: *scan line decimation* and *vertical filtering*.

Scan Line Decimation

The easiest approach is to throw away every other active scan line in each noninterlaced frame. Although the cost is minimal, there are problems with this approach, especially with the top and bottom of objects.

If there is a sharp vertical transition of color or intensity, it will flicker at one-half the frame rate. The reason is that it is only displayed every other field as a result of the decimation. For example, a horizontal line that is one noninterlaced scan line wide will flicker on and off. Horizontal lines that are two noninterlaced scan lines wide will oscillate up and down.

ALERT!

Simple decimation may also add aliasing artifacts. While not necessarily visible, they will affect any future processing of the picture.

Vertical Filtering

A better solution is to use two or more lines of noninterlaced data to generate one line of interlaced data. Fast vertical transitions are smoothed out over several interlaced lines.

For a 3-line filter, typical coefficients are [0.25, 0.5, 0.25]. Using more than three lines usually results in excessive blurring, making small text difficult to read.

An alternate implementation uses IIR rather than FIR filtering. In addition to averaging, this technique produces a reduction in brightness around objects, further reducing flicker.

Note that care must be taken at the beginning and end of each frame in the event that fewer scan lines are available for filtering.

INTERLACED-TO-NONINTERLACED CONVERSION

In some applications, it is necessary to display an interlaced video signal on a noninterlaced display. Thus, some form of deinterlacing or progressive scan conversion may be required.

ALERT!

Note that deinterlacing must be performed on component video signals (such as R′G′B′ or YCbCr). Composite color video signals (such as NTSC or PAL) cannot be deinterlaced directly due to the presence of color subcarrier phase information, which would be meaningless after processing. These signals must be decoded into component color signals, such as R′G′B′ or YCbCr, prior to deinterlacing.

There are two fundamental deinterlacing algorithms: video mode and film mode. Video mode deinterlacing can be further broken down into inter-field and intra-field processing.

The goal of a good deinterlacer is to correctly choose the best algorithm needed at a particular moment. In systems where the vertical resolution of the source and display do not match (due to, for example, displaying SDTV content on an HDTV), the deinterlacing and vertical scaling can be merged into a single process.

Video Mode: Intra-Field Processing

This is the simplest method for generating additional scan lines using only information in the original field. The computer industry has coined this technique as bob.

Although there are two common techniques for implementing intra-field processing, scan line duplication and scan line interpolation, the resulting vertical resolution is always limited by the content of the original field.

Scan Line Duplication

Scan line duplication simply duplicates the previous active scan line. Although the number of active scan lines is doubled, there is no increase in the vertical resolution.

Scan Line Interpolation

Scan line interpolation generates interpolated scan lines between the original active scan lines. Although the number of active scan lines is doubled, the vertical resolution is not.

The simplest implementation uses linear interpolation to generate a new scan line between two input scan lines. Better results, at additional cost, may be achieved by using a FIR filter:

Fractional Ratio Interpolation

In many cases, there is a periodic, but non-integral, relationship between the number of input scan lines and the number of output scan lines. In this case, fractional ratio interpolation may be necessary, similar to the polyphase filtering used for scaling only performed in the vertical direction. This technique combines deinterlacing and vertical scaling into a single process.

Variable Interpolation

In a few cases, there is no periodicity in the relationship between the number of input and output scan lines. Therefore, in theory, an infinite number of filter phases and coefficients are required. Since this is not feasible, the solution is to use a large, but finite, number of filter phases. The number of filter phases determines the interpolation accuracy. This technique also combines deinterlacing and vertical scaling into a single process.

Video Mode: Inter-Field Processing

In this method, video information from more than one field is used to generate a single progressive frame. This method can provide higher vertical resolution since it uses content from more than a single field.

Field Merging

This technique merges two consecutive fields together to produce a frame of video. At each field time, the active scan lines of that field are merged with the active scan lines of the previous field. The result is that for each input field time, a pair of fields combine to generate a frame. Although simple to implement, the vertical resolution is doubled only in regions of no movement.

Moving objects will have artifacts, also called combing, due to the time difference between two fields—a moving object is located in a different position from one field to the next. When the two fields are merged, moving objects will have a double image.

It is common to soften the image slightly in the vertical direction to attempt to reduce the visibility of combing. When implemented, it causes a loss of vertical resolution and jitter on movement and pans.

Insider Info

The computer industry refers to this technique as weave, *but weave also includes the inverse telecine process to remove any 3:2 pull-down present in the source. Theoretically, this eliminates the double image artifacts since two identical fields are now being merged.*

Motion Adaptive Deinterlacing

A good deinterlacing solution is to use field merging for still areas of the picture and scan line interpolation for areas of movement. To accomplish this, motion, on a sample-by-sample basis, must be detected over the entire picture in real time, requiring processing several fields of video.

As two fields are combined, full vertical resolution is maintained in still areas of the picture, where the eye is most sensitive to detail. The sample differences may have any value, from 0 (no movement and noise-free) to maximum (for example, a change from full intensity to black). A choice must be made when to use a sample from the previous field (which is in the wrong location due to motion) or to interpolate a new sample from adjacent scan lines in the current field. Sudden switching between methods is visible, so crossfading (also called soft switching) is used. At some magnitude of sample difference, the loss of resolution due to a double image is equal to the loss of resolution due to interpolation. That amount of motion should result in the crossfader being at the 50% point. Less motion will result in a fade towards field merging and more motion in a fade towards the interpolated values.

Insider Info

Rather than "per pixel" motion adaptive deinterlacing, which makes decisions for every sample, some low-cost solutions use "per field" motion adaptive deinterlacing. In this case, the algorithm is selected each field, based on the amount of motion between the fields. "Per pixel" motion adaptive deinterlacing, although difficult to implement, looks quite good when properly done. "Per field" motion adaptive deinterlacing rarely looks much better than vertical interpolation.

Motion-Compensated Deinterlacing

Motion-compensated (or motion vector steered) deinterlacing is several orders of magnitude more complex than motion adaptive deinterlacing, and is commonly found in pro-video format converters.

Motion-compensated processing requires calculating motion vectors between fields for each sample, and interpolating along each sample's motion trajectory. Motion vectors must also be found that pass through each of any missing samples. Areas of the picture may be covered or uncovered as you move between frames. The motion vectors must also have sub-pixel accuracy, and be determined in two temporal directions between frames.

The motion vector errors used by MPEG are self-correcting since the residual difference between the predicted macroblocks is encoded. As motion-compensated deinterlacing is a single-ended system, motion vector errors will produce artifacts, so different search and verification algorithms must be used.

DCT-BASED COMPRESSION

The transform process of many video compression standards is based on the Discrete Cosine Transform, or DCT. The easiest way to envision it is as a filter bank with all the filters computed in parallel.

During encoding, the DCT is usually followed by several other operations, such as quantization, zig-zag scanning, run-length encoding, and variable-length encoding. During decoding, this process flow is reversed.

Many times, the terms macroblocks and blocks are used when discussing video compression. Figure 5.19 illustrates the relationship between these two terms, and shows why transform processing is usually done on 8×8 samples.

DCT

The 8×8 DCT processes an 8×8 block of samples to generate an 8×8 block of DCT coefficients, as shown in Figure 5.20. The input may be samples

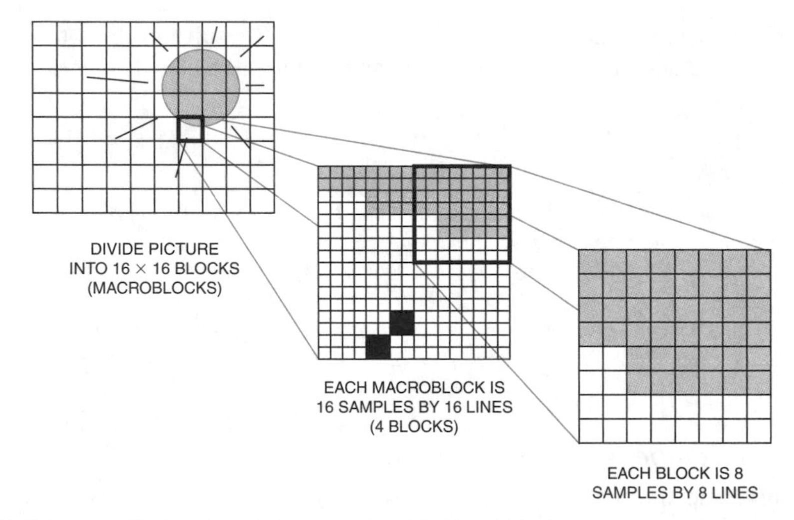

FIGURE 5.19 The Relationship between Macroblocks and Blocks.

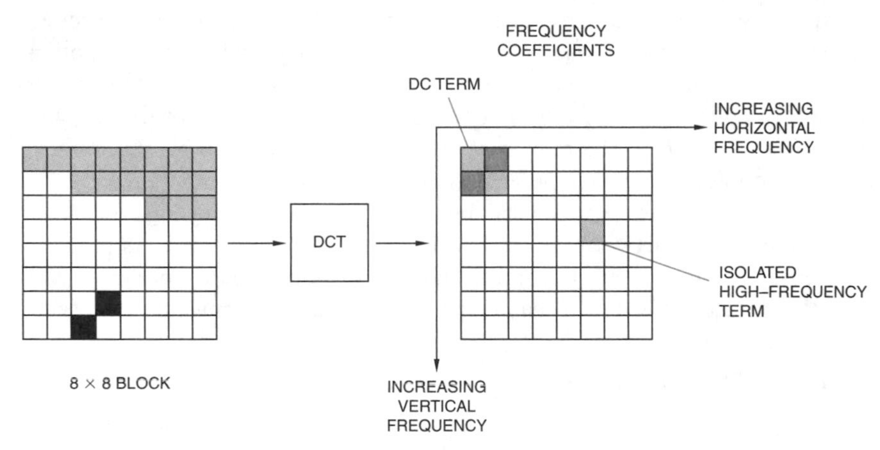

FIGURE 5.20 The DCT Processes the 8 × ?8 Block of Samples or Error Terms to Generate an 8 × ?8 Block of DCT Coefficients.

from an actual frame of video or motion-compensated difference (error) values, depending on the encoder mode of operation. Each DCT coefficient indicates the amount of a particular horizontal or vertical frequency within the block.

DCT coefficient (0,0) is the DC coefficient, or average sample value. Since natural images tend to vary only slightly from sample to sample, low frequency coefficients are typically larger values and high frequency coefficients are typically smaller values.

$$F(u,v) = 0.25C(u)C(v)\sum_{x=0}^{7}\sum_{y=0}^{7} f(x,y)\cos(((2x+1)u\pi)/16)\cos(((2y+1)v\pi)/16)$$

u, v, x, y = 0, 1, 2, . . ., 7
(x, y) are spatial coordinates in the sample domain
(u, v) are coordinates in the transform domain

FIGURE 5.21 8 × 8 Two-Dimensional DCT Definition.

The 8 × 8 DCT is defined in Figure 5.21. f(x, y) denotes sample (x, y) of the 8 × 8 input block and F(u,v) denotes coefficient (u, v) of the DCT transformed block.

A reconstructed 8 × 8 block of samples is generated using an 8 × 8 inverse DCT (IDCT). Although exact reconstruction is theoretically achievable, it is not practical duc to finitc-prccision arithmctic, quantization and differing IDCT implementations. As a result, there are mismatches between different IDCT implementations.

Mismatch control attempts to reduce the drift between encoder and decoder IDCT results by eliminating bit patterns having the greatest contribution towards mismatches.

MPEG-1 mismatch control is known as "oddification" since it forces all quantized DCT coefficients to negative values. MPEG-2 and MPEG-4.2 use an improved method called "LSB toggling" which affects only the LSB of the 63rd DCT coefficient after inverse quantization.

H.264 (also known as MPEG-4.10) neatly sidesteps the issue by using an "exact-match inverse transform." Every decoder will produce exactly the same pictures, all else being equal.

Quantization

The 8 × 8 block of DCT coefficients is quantized, which reduces the overall precision of the integer coefficients and tends to eliminate high frequency coefficients, while maintaining perceptual quality. Higher frequencies are usually quantized more coarsely (fewer values allowed) than lower frequencies, due to visual perception of quantization error. The quantizer is also used for constant bit-rate

Zig-Zag Scanning

The quantized DCT coefficients are rearranged into a linear stream by scanning them in a zig-zag order. This rearrangement places the DC coefficient first, followed by frequency coefficients arranged in order of increasing frequency. This produces long runs of zero coefficients.

Run Length Coding

The linear stream of quantized frequency coefficients is converted into a series of [run, amplitude] pairs. [run] indicates the number of zero coefficients, and [amplitude] the nonzero coefficient that ended the run.

Variable-Length Coding

The [run, amplitude] pairs are coded using a variable-length code, resulting in additional lossless compression. This produces shorter codes for common pairs and longer codes for less common pairs.

This coding method produces a more compact representation of the DCT coefficients, as a large number of DCT coefficients are usually quantized to zero and the re-ordering results (ideally) in the grouping of long runs of consecutive zero values.

INSTANT SUMMARY

This chapter covers various types of video processing, including:

- Display enhancement
- Video mixing and graphics overlay
- Chroma keying
- Video scaling
- Scan rate conversion
- Noninterlaced to interlaced conversion
- Interlaced to noninterlaced conversion

Chapter 6

NTSC, PAL, and SECAM

In an Instant

- Definitions
- NTSC overview
- PAL overview
- SECAM overview
- Enhanced television programming

Definitions

The three major world color television standards are NTSC, PAL, and SECAM. In this chapter we will examine some features of all three. First we'll define some terms used in the rest of the chapter.

Hue, saturation and *luminance* are terms that describe color. Hue refers to the wavelength of the color, which means that hue is the term used for the base color—red, green, yellow, etc. Hue is completely separate from the saturation or intensity of the color. For example, a red hue could look brown at low saturation, bright red at a higher level of saturation, or pink at a high brightness level.

A *subcarrier* is a secondary signal containing additional information that is added to a main signal.

A *modulator* is a circuit that combines two different signals in such a way that they can be pulled apart later and the information obtained. For example, the NTSC video system may use the YIQ or YUV color space, with the I and Q or U and V signals containing all of the color information for the picture. Two 3.58 MHz color subcarriers (90 degrees out of phase) are modulated by the I and Q or U and V components and added together to create the chroma part of the NTSC video.

A *color burst* is an analog waveform of a specific frequency and amplitude that is positioned between the trailing edge of horizontal sync and the start of active video. The color burst tells the NTSC or PAL video decoder how to decode the color information contained in that line of active video.

NTSC OVERVIEW

The first color television system was developed in the United States, and on December 17, 1953, the Federal Communications Commission (FCC) approved the transmission standard, with broadcasting approved to begin January 23, 1954. Most of the work for developing a color transmission standard that was compatible with the (then current) 525-line, 60-field-per-second, 2:1interlaced monochrome standard was done by the National Television System Committee (NTSC).

Luminance Information

The monochrome luminance (Y) signal is derived from gamma-corrected red, green, and blue (R′G′B′) signals:

$$Y = 0.299R' + 0.587G' + 0.114B'$$

Technology Trade-offs

Due to the sound subcarrier at 4.5 MHz, a requirement was made that the color signal fit within the same bandwidth as the monochrome video signal (0–4.2 MHz). For economic reasons, another requirement was made that monochrome receivers must be able to display the black and white portion of a color broadcast and that color receivers must be able to display a monochrome broadcast.

Color Information

Insider Info

The eye is most sensitive to spatial and temporal variations in luminance; therefore, luminance information was still allowed the entire bandwidth available (0–4.2 MHz). Color information, which the eye is less sensitive and which therefore requires less bandwidth, is represented as hue and saturation information.

The hue and saturation information is transmitted using a 3.58-MHz subcarrier, encoded so that the receiver can separate the hue, saturation, and luminance information and convert them back to RGB signals for display. Although this allows the transmission of color signals within the same bandwidth as monochrome signals, the problem still remains as to how to separate the color and luminance information cost-effectively, since they occupy the same portion of the frequency spectrum.

To transmit color information, U and V or I and Q "color difference" signals are used:

$$R' - Y = 0.701R' - 0.587G' - 0.114B'$$
$$B' - Y = -0.299R' - 0.587G' + 0.866B'$$

$$U = 0.492(B' - Y)$$
$$V = 0.877(R' - Y)$$

$$\begin{aligned} I &= 0.212R' - 0.523G' + 0.311B' \\ &= V\sin 33° + U\cos 33° \\ &= 0.736(R' - Y) + 0.413(B' - Y) \end{aligned}$$

$$\begin{aligned} Q &= 0.212R' - 0.523G' + 0.311B' \\ &= V\sin 33° + U\cos 33° \\ &= 0.478(R' - Y) + 0.413(B' - Y) \end{aligned}$$

The scaling factors to generate U and V from (B′ − Y) and (R′ − Y) were derived due to overmodulation considerations during transmission. If the full range of (B′ − Y) and (R′ − Y) were used, the modulated chrominance levels would exceed what the monochrome transmitters were capable of supporting. Experimentation determined that modulated subcarrier amplitudes of 20% of the Y signal amplitude could be permitted above white and below black. The scaling factors were then selected so that the maximum level of 75% color would be at the white level.

I and Q were initially selected since they more closely related to the variation of color acuity than U and V. The color response of the eye decreases as the size of viewed objects decreases. Small objects, occupying frequencies of 1.3–2.0 MHz, provide little color sensation. Medium objects, occupying the 0.6–1.3 MHz frequency range, are acceptable if reproduced along the orange-cyan axis. Larger objects, occupying the 0–0.6 MHz frequency range, require full three-color reproduction.

The I and Q bandwidths were chosen accordingly, and the preferred color reproduction axis was obtained by rotating the U and V axes by 33°. The Q component, representing the green-purple color axis, was band-limited to about 0.6 MHz. The I component, representing the orange-cyan color axis, was band-limited to about 1.3 MHz.

Another advantage of limiting the I and Q bandwidths to 1.3 MHz and 0.6 MHz, respectively, is to minimize crosstalk due to asymmetrical sidebands as a result of lowpass filtering the composite video signal to about 4.2 MHz. Q is a double sideband signal; however, I is asymmetrical, bringing up the possibility of crosstalk between I and Q. The symmetry of Q avoids crosstalk into I; since Q is bandwidth limited to 0.6 MHz, I crosstalk falls outside the Q bandwidth.

U and V, both bandwidth-limited to 1.3 MHz, are now commonly used instead of I and Q. When broadcast, UV crosstalk occurs above 0.6 MHz; however, this is not usually visible due to the limited UV bandwidths used by NTSC decoders for consumer equipment.

The UV and IQ vector diagram is shown in Figure 6.1.

Color Modulation

I and Q (or U and V) are used to modulate a 3.58 MHz color subcarrier using two balanced modulators operating in phase quadrature: one modulator is

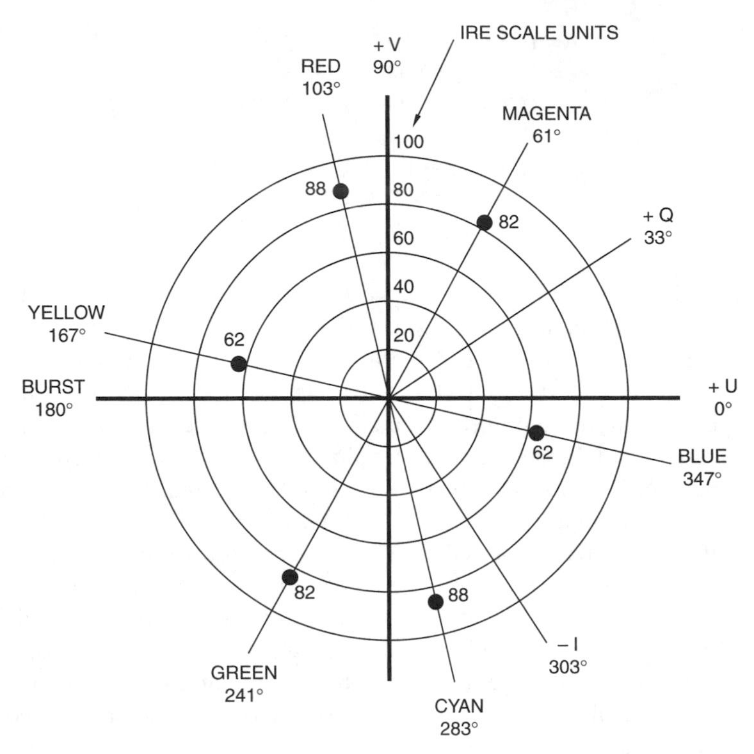

FIGURE 6.1 UV and IQ Vector Diagram for 75% Color Bars.

driven by the subcarrier at sine phase; the other modulator is driven by the subcarrier at cosine phase.

Hue information is conveyed by the chrominance phase relative to the subcarrier. Saturation information is conveyed by chrominance amplitude. In addition, if an object has no color (such as a white, gray, or black object), the subcarrier is suppressed.

Composite Video Generation

The modulated chrominance is added to the luminance information along with appropriate horizontal and vertical sync signals, blanking information, and color burst information, to generate the composite color video waveform shown in Figure 6.2.

The I and Q (or U and V) information can be transmitted without loss of identity as long as the proper color subcarrier phase relationship is maintained at the encoding and decoding process. A *color burst* signal, consisting of nine cycles of the subcarrier frequency at a specific phase, follows most horizontal sync pulses, and provides the decoder a reference signal so as to be able to recover the I and Q (or U and V) signals properly.

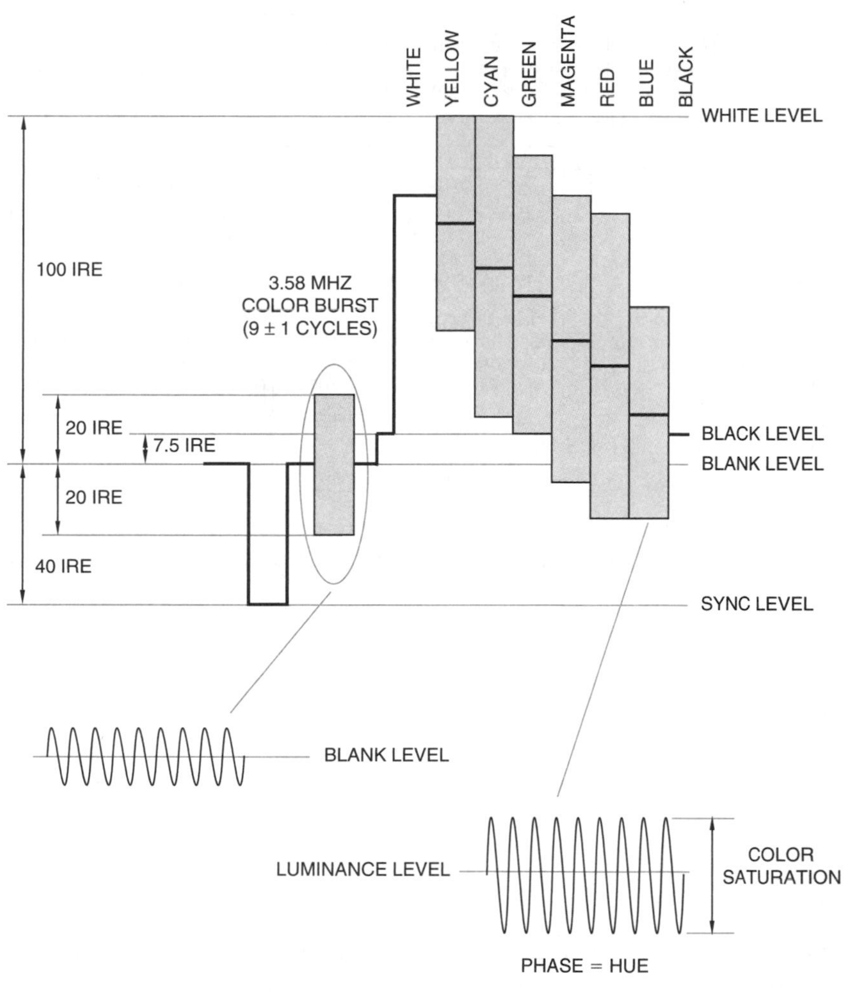

FIGURE 6.2 (M) NTSC Composite Video Signal for 75% Color Bars.

NTSC Standards

Figure 6.3 shows the common designations for NTSC systems. The letter M refers to the monochrome standard for line and field rates (525/59.94), a video bandwidth of 4.2 MHz, an audio carrier frequency 4.5 MHz above the video carrier frequency, and an RF channel bandwidth of 6 MHz. NTSC refers to the technique to add color information to the monochrome signal.

NTSC 4.43 is commonly used for multi-standard analog VCRs. The horizontal and vertical timing is the same as (M) NTSC; color encoding uses the PAL modulation format and a 4.43361875-MHz color subcarrier frequency.

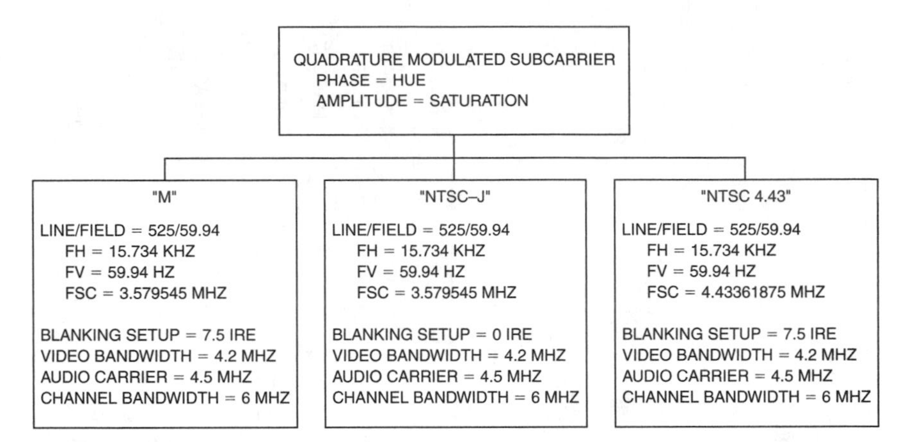

FIGURE 6.3 Common NTSC Systems.

Noninterlaced NTSC is a 262-line, 60 frames-per-second version of NTSC. This format is identical to standard (M) NTSC, except that there are 262 lines per frame.

Insider Info

NTSC–J, used in Japan, is the same as (M) NTSC, except there is no blanking pedestal during active video. Thus, active video has a nominal amplitude of 714 mV.

PAL OVERVIEW

Europe delayed adopting a color television standard, evaluating various systems between 1953 and 1967 that were compatible with their 625-line, 50-field-per-second, 2:1 interlaced monochrome standard. The NTSC specification was modified to overcome the high order of phase and amplitude integrity required during broadcast to avoid color distortion. The Phase Alternation Line (PAL) system implements a line-by-line reversal of the phase of one of the color components, originally relying on the eye to average any color distortions to the correct color. Broadcasting began in 1967 in Germany and the United Kingdom, with each using a slightly different variant of the PAL system.

Luminance Information

The monochrome luminance (Y) signal is derived from R′G′B′:

$$Y = 0.299R' + 0.587G' + 0.114B'$$

As with NTSC, the luminance signal occupies the entire video bandwidth. PAL has several variations, depending on the video bandwidth and placement of the audio subcarrier. The composite video signal has a bandwidth of 4.2, 5.0, 5.5, or 6.0 MHz, depending on the specific PAL standard.

Color Information

To transmit color information, U and V are used:

$$U = 0.492(B' - Y)$$
$$V = 0.877(R' - Y)$$

U and V have a typical bandwidth of 1.3 MHz.

Color Modulation

As in the NTSC system, U and V are used to modulate the color subcarrier using two balanced modulators operating in phase quadrature: one modulator is driven by the subcarrier at sine phase; the other modulator is driven by the subcarrier at cosine phase. The outputs of the modulators are added together to form the modulated chrominance signal:

$$C = U \sin \omega t \pm V$$
$$\omega = 2\pi F_{SC}$$

$$F_{SC} = 4.43361875\,\text{MHz}(\pm 5\,\text{Hz}) \text{ for (B, D, G, H, I, N) PAL}$$
$$F_{SC} = 3.58205625\,\text{MHz}(\pm 5\,\text{Hz}) \text{ for } (N_C)\text{PAL}$$
$$F_{SC} = 3.57561143\,\text{MHz}(\pm 10\,\text{Hz}) \text{ for (M)PAL}$$

In PAL, the phase of V is reversed every other line. V was chosen for the reversal process since it has a lower gain factor than U and therefore is less susceptible to a one-half FH switching rate imbalance. The result of alternating the V phase at the line rate is that any color subcarrier phase errors produce complementary errors, allowing line-to-line averaging at the receiver to cancel the errors and generate the correct hue with slightly reduced saturation. This technique requires the PAL receiver to be able to determine the correct V phase. This is done using a technique known as *AB sync*, *PAL sync*, *PAL switch*, or *swinging burst*, consisting of alternating the phase of the color burst by ±45° at the line rate.

Technology Trade-offs

Simple PAL decoders rely on the eye to average the line-by-line hue errors. Standard PAL decoders use a 1 H delay line to separate U from V in an averaging process. Both implementations have the problem of Hanover bars, in which pairs of adjacent lines have a real and complementary hue error. Chrominance vertical resolution is reduced as a result of the line averaging process.

Composite Video Generation

The modulated chrominance is added to the luminance information along with appropriate horizontal and vertical sync signals, blanking signals, and color burst signals, to generate the composite color video waveform shown in Figure 6.4.

Like NTSC, the luminance components are spaced at FH intervals due to horizontal blanking. Since the V component is switched symmetrically at one-half the line rate, only odd harmonics are generated, resulting in V components

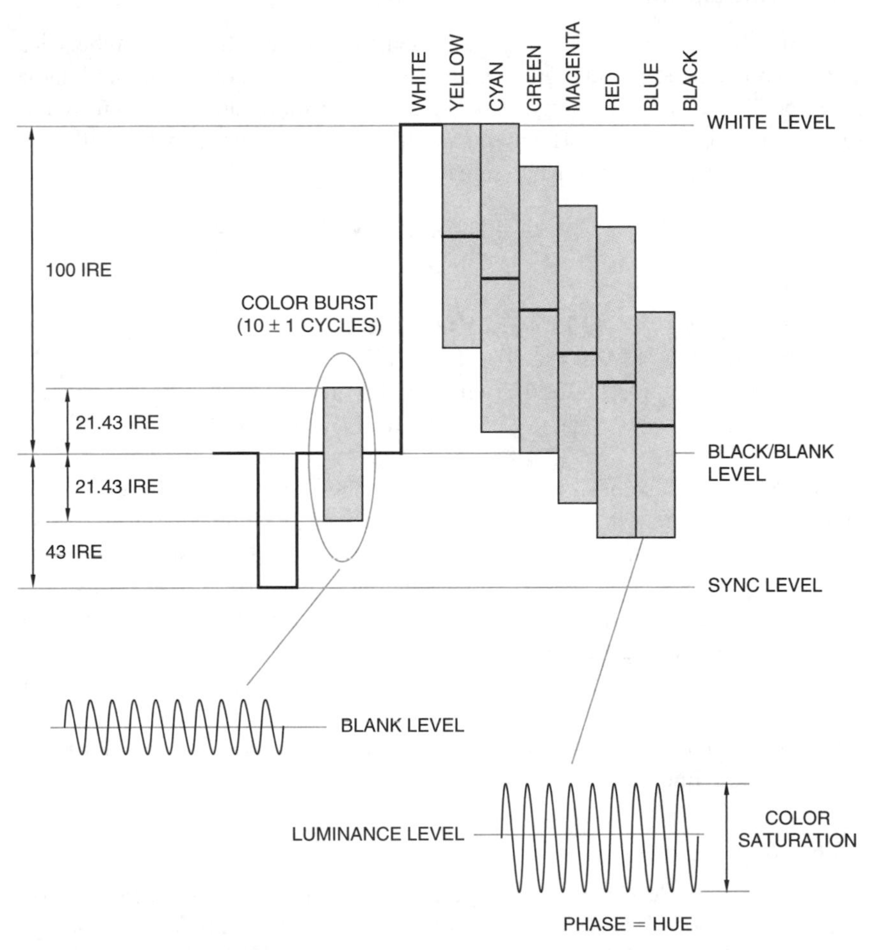

FIGURE 6.4 (B, D, G, H, I, N_C) PAL Composite Video Signal for 75% Color Bars.

that are spaced at intervals of FH. The V components are spaced at half-line intervals from the U components, which also have FH spacing. If the subcarrier had a half-line offset like NTSC uses, the U components would be perfectly interleaved, but the V components would coincide with the Y components and thus not be interleaved, creating vertical stationary dot patterns. For this reason, PAL uses a 1/4 line offset for the subcarrier frequency:

PAL Standards

Figure 6.5 shows the common designations for PAL systems. The letters refer to the monochrome standard for line and field rate, video bandwidth (4.2, 5.0, 5.5, or 6.0 MHz), audio carrier relative frequency, and RF channel bandwidth (6.0, 7.0, or 8.0 MHz). PAL refers to the technique to add color information to the monochrome signal. Noninterlaced PAL is a 312-line, 50-frames-per-second version of PAL common among video games and on-screen displays. This format is identical to standard PAL, except that there are 312 lines per frame.

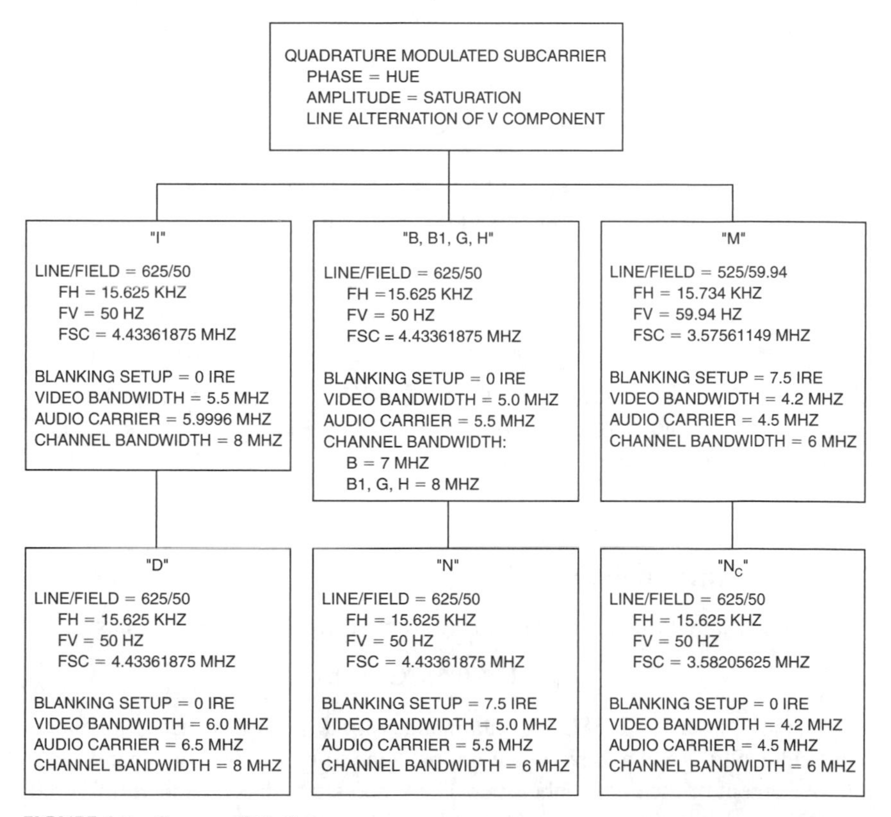

FIGURE 6.5 Common PAL Systems.

PALplus

PALplus (ITU-R BT.1197 and ETSI ETS 300 731) is the result of a cooperative project started in 1990, undertaken by several European broadcasters. By 1995, they wanted to provide an enhanced definition television system (EDTV), compatible with existing receivers. PALplus has been transmitted by a few broadcasters since 1994.

A PALplus picture has a 16:9 aspect ratio. On conventional TVs, it is displayed as a 16:9 letterboxed image with 430 active lines. On PALplus TVs, it is displayed as a 16:9 picture with 574 active lines, with extended vertical resolution. The full video bandwidth is available for luminance detail. Cross color artifacts are reduced by clean encoding.

A PALplus TV has the option of deinterlacing a film mode signal and displaying it on a 50-Hz progressive-scan display or using field repeating on a 100-Hz interlaced display.

SECAM OVERVIEW

SECAM (Sequentiel Couleur Avec Mémoire, or Sequential Color with Memory) was developed in France (broadcasting started in 1967) due to the realization that, if color could be bandwidth-limited horizontally, why not also vertically? The two pieces of color information (Db and Dr) added to the monochrome signal could be transmitted on alternate lines, avoiding the possibility of crosstalk.

The receiver requires memory to store one line so that it is concurrent with the next line, and also requires the addition of a line-switching identification technique.

Insider Info

Like PAL, SECAM is a 625-line, 50-field-per-second, 2:1 interlaced system. SECAM was adopted by other countries; however, many are changing to PAL due to the abundance of professional and consumer PAL equipment.

Luminance Information

The monochrome luminance (Y) signal is derived from (R′G′B′) signals:

$$Y = 0.299R' + 0.587G' + 0.114B'$$

As with NTSC and PAL, the luminance signal occupies the entire video bandwidth. SECAM has several variations, depending on the video bandwidth and placement of the audio subcarrier. The video signal has a bandwidth of 5.0 or 6.0 MHz, depending on the specific SECAM standard.

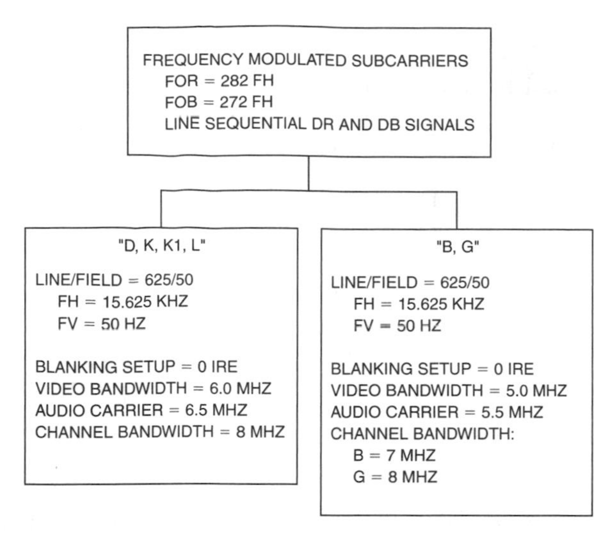

FIGURE 6.6 Common SECAM Systems.

The SECAM refers to the technique to add color information to the monochrome signal.

ENHANCED TELEVISION PROGRAMMING

The enhanced television programming standard (SMPTE 363M) is used for creating and delivering enhanced and interactive programs. The enhanced content can be delivered over a variety of mediums—including analog and digital television broadcasts—using terrestrial, cable, and satellite networks. In defining how to create enhanced content, the specification defines the minimum receiver functionality. To minimize the creation of new specifications, it leverages Internet technologies such as HTML and Java-script. The benefits of doing this are that there are already millions of pages of potential content, and the ability to use existing web-authoring tools.

The specification mandates that receivers support, as a minimum, HTML 4.0, Javascript 1.1, and Cascading Style Sheets. Supporting additional capabilities, such as Java and VRML, is optional. This ensures content is available to the maximum number of viewers.

For increased capability, a new “tv:” attribute is added to the HTML. This attribute enables the insertion of the television program into the content, and may be used in an HTML document anywhere that a regular image may be placed. Creating an enhanced content page that displays the current television channel anywhere on the display is as easy as inserting an image in an HTML document.

The specification also defines how the receivers obtain the content and how they are informed that enhancements are available. The latter task is accomplished with triggers.

Color Information

SECAM transmits Db information during one line and Dr information du the next line; luminance information is transmitted each line. Db and Dı scaled versions of B′ − Y and R′ − Y:

$$Dr = -1.902(R' - Y)$$
$$Db = 1.505(B' - Y)$$

Since there is an odd number of lines, any given line contains Db inform tion on one field and Dr information on the next field. The decoder requires 1 H delay, switched synchronously with the Db and Dr switching, so that D and Dr exist simultaneously in order to convert to YCbCr or RGB.

Color Modulation

SECAM uses FM modulation to transmit the Db and Dr color difference information, with each component having its own subcarrier.

Db and Dr are lowpass filtered to 1.3 MHz and pre-emphasis is applied. After pre-emphasis, Db and Dr frequency modulate their respective subcarriers. The frequencies of the subcarriers represent no color information. The choice of frequency shifts reflects the idea of keeping the frequencies representing critical colors away from the upper limit of the spectrum to minimize distortion.

After modulation of Db and Dr, subcarrier pre-emphasis is applied, changing the amplitude of the subcarrier as a function of the frequency deviation. The intention is to reduce the visibility of the subcarriers in areas of low luminance and to improve the signal-to-noise ratio of highly saturated colors. Db and Dr information is transmitted on alternate scan lines. Note that subcarrier phase information in the SECAM system carries no picture information.

Composite Video Generation

The subcarrier data is added to the luminance along with appropriate horizontal and vertical sync signals, blanking signals, and burst signals to generate composite video.

As with PAL, SECAM requires some means of identifying the line-switching sequence. Modern practice has been to use an FOR/FOB burst after most horizontal syncs to derive the switching synchronization information.

SECAM Standards

Figure 6.6 shows the common designations for SECAM systems. The letters refer to the monochrome standard for line and field rates, video bandwidth (5.0 or 6.0 MHz), audio carrier relative frequency, and RF channel bandwidth.

Triggers

Triggers alert receivers to content enhancements, and contain information about the enhancements. Among other things, triggers contain a universal resource locator (URL) that defines the location of the enhanced content. Content may reside locally—such as when delivered over the network and cached to a local hard drive—or it may reside on the Internet or another network.

Triggers may also contain a human-readable description of the content. For example, it may contain the description "Press ORDER to order this product," which can be displayed for the viewer. Triggers also may contain expiration information, indicating how long the enhancement should be offered to the viewer.

Lastly, triggers may contain scripts that trigger the execution of Javascript within the associated HTML page, to support synchronization of the enhanced content with the video signal and updating of dynamic screen data.

The processing of triggers is defined in SMPTE 363M and is independent of the method used to carry them.

Transports

Besides defining how content is displayed and how the receiver is notified of new content, the specification also defines how content is delivered. Because a receiver may not have an Internet connection, the specification describes two models for delivering content. These two models are called transports, and the two transports are referred to as Transport Type A and Transport Type B.

If the receiver has a back-channel (or return path) to the Internet, Transport Type A will broadcast the trigger and the content will be pulled over the Internet.

If the receiver does not have an Internet connection, Transport Type B provides for delivery of both triggers and content via the broadcast medium. Announcements are sent over the network to associate triggers with content streams. An announcement describes the content, and may include information regarding bandwidth, storage requirements, and language.

INSTANT SUMMARY

This chapter covered features of NTSC, PAL and SECAM television standards, including:

- Color and luminance information
- Color modulation
- Composite video generation
- Common standards designations
- Enhanced television programming

Chapter 7

MPEG-1, MPEG-2, MPEG-4, and H.264

In an Instant

- MPEG Definitions
- MPEG-1
- MPEG-2
- MPEG-4
- H.264

MPEG Definitions

MPEG stands for Motion Picture Experts Group, an ISO/IEC standards organization group that is developing various compression algorithms. In this chapter we will take a look at the most commonly used MPEG standards for video compression. First we will define some terms.

MPEG-1 was the first MPEG standard defining the compression format for real-time audio and video. Features include random access, fast forward, and reverse playback.

MPEG-2 extends the MPEG-1 standard to cover a wider range of applications. Higher video resolutions are supported, to allow for HDTV applications, and both progressive and interlaced video are supported.

MPEG-4 supports an object-based approach, where scenes are modeled as compositions of objects, both natural and synthetic, with which the user can interact.

H-264, a next-generation video codec, is included in the MPEG-4 standard as Part 10.

Lossless compression is a term that refers to the case when the compressed data is still exactly the same as the original data—no information is lost in the compression. Conversely, *lossy* compression is the exact opposite of lossless. The regenerated data is different from the original data (the differences may or may not be noticeable).

MPEG-1

MPEG-1 was developed specifically for storing and distributing audio and video. It is used as the basis for the original video CDs (VCD).

Insider Info

The channel bandwidth and image resolution were set by the available media at the time (CDs). The goal was playback of digital audio and video using a standard compact disc with a bit-rate of 1.416 Mbps (1.15 Mbps of this is for video).

MPEG-1 is an ISO standard (ISO/IEC 11172), and consists of six parts:

System	ISO/IEC 11172–1
Video	ISO/IEC 11172–2
Audio	ISO/IEC 11172–3
Low bit-rate audio	ISO/IEC 13818–3
Conformance testing	ISO/IEC 11172–4
Simulation software	ISO/IEC 11172–5

The bitstreams implicitly define the decompression algorithms. The compression algorithms are up to the individual manufacturers, allowing a proprietary advantage to be obtained within the scope of an international standard.

MPEG vs. JPEG

JPEG (ISO/IEC 10918-Joint Photographic Experts Group) is an image compression standard that was designed primarily for still images and single video frames. It doesn't handle bi-level (black and white) images efficiently, and pseudo-color images have to be expanded into the unmapped color representation prior to processing. JPEG images may be of any resolution and color space, with both lossy and lossless algorithms available. If you run JPEG fast enough, you can compress motion video. This is called motion JPEG, or M-JPEG.

Since JPEG is such a general-purpose standard, it has many features and capabilities. By adjusting the various parameters, compressed image size can be traded against reconstructed image quality over a wide range. Image quality ranges from "browsing" (100:1 compression ratio) to "indistinguishable from the source" (about 3:1 compression ratio). Typically, the threshold of visible difference between the source and reconstructed images is somewhere between a 10:1 to 20:1 compression ratio.

How It Works

JPEG does not use a single algorithm, but rather a family of four, each designed for a certain application. The most familiar lossy algorithm is sequential DCT. *Either Huffman encoding (baseline JPEG) or arithmetic encoding may be used. When the image is decoded, it is decoded left-to-right, top-to-bottom.* Progressive DCT *is another lossy algorithm, requiring multiple scans of the image. When the image is decoded, a coarse approximation of the full image is available right away, with the quality progressively improving until complete. This makes it ideal for applications such as image database browsing. Either spectral selection, successive approximation, or both may be used. The spectral selection option encodes the lower-frequency DCT coefficients first (to obtain an image quickly), followed by the higher-frequency ones (to add more detail). The successive approximation option encodes the more significant bits of the DCT coefficients first, followed by the less significant bits.*

The hierarchical mode represents an image at multiple resolutions. For example, there could be 512 × 512, × 1024 × 1024, and 2048 × 2048 versions of the image. Higher-resolution images are coded as differences from the next smaller image, requiring fewer bits than they would if stored independently. Of course, the total number of bits is greater than that needed to store just the highest-resolution image. Note that the individual images in a hierarchical sequence may be coded progressively if desired.

Also supported is a lossless spatial algorithm that operates in the pixel domain as opposed to the transform domain. A prediction is made of a sample value using up to three neighboring samples. This prediction then is subtracted from the actual value and the difference is losslessly coded using either Huffman or arithmetic coding. Lossless operation achieves about a 2:1 compression ratio.

Technology Trade-offs

Since video is just a series of still images, and baseline JPEG encoders and decoders were readily available, people used baseline JPEG to compress real-time video (also called motion JPEG or MJPEG). However, this technique does not take advantage of the frame-to-frame redundancies to improve compression, as does MPEG. JPEG is *symmetrical*, meaning the cost of encoding and decoding is roughly the same. MPEG, on the other hand, was designed primarily for mastering a video once and playing it back many times on many platforms. To minimize the cost of MPEG hardware decoders, MPEG was designed to be asymmetrical, with the encoding process requiring about 100× the computing power of the decoding process.

Since MPEG is targeted for specific applications, the hardware usually supports only a few specific resolutions. Also, only one color space (YCbCr) is supported using 8-bit samples. MPEG is also optimized for a limited range of compression ratios.

If capturing video for editing, you can use either baseline JPEG or I-frame-only (intra-frame) MPEG to compress to disc in real-time. Using JPEG requires that the system be able to transfer data and access the hard disk at bit-rates of about 4Mbps for SIF (Standard Input Format) resolution. Once the editing is done, the result can be converted into MPEG for maximum compression.

Quality Issues

At bit-rates of about 3–4Mbps, "broadcast quality" is achievable with MPEG-1. However, sequences with complex spatial-temporal activity (such as sports) may require up to 5–6Mbps due to the frame-based processing of MPEG-1. MPEG-2 allows similar "broadcast quality" at bit-rates of about 4–6Mbps by supporting field-based processing.

Several factors affect the quality of MPEG-compressed video:

- the resolution of the original video source
- the bit-rate (channel bandwidth) allowed after compression
- motion estimator effectiveness

One limitation of the quality of the compressed video is determined by the resolution of the original video source. If the original resolution was too low, there will be a general lack of detail.

Motion estimator effectiveness determines motion artifacts, such as a reduction in video quality when movement starts or when the amount of movement is above a certain threshold. Poor motion estimation will contribute to a general degradation of video quality.

Most importantly, the higher the bit-rate (channel bandwidth), the more information that can be transmitted, allowing fewer motion artifacts to be present or a higher-resolution image to be displayed. Generally speaking, decreasing the bit-rate does not result in a graceful degradation of the decoded video quality. The video quality rapidly degrades, with the 8×8 blocks becoming clearly visible once the bit-rate drops below a given threshold.

Audio Overview

MPEG-1 uses a family of three audio coding schemes, called Layer I, Layer II, and Layer III, with increasing complexity and sound quality. The three layers are hierarchical: a Layer III decoder handles Layers I, II, and III; a Layer II decoder handles only Layers I and II; a Layer I decoder handles only Layer I. All layers support 16-bit audio using 16, 22.05, 24, 32, 44.1, or 48kHz sampling rates.

For each layer, the bitstream format and the decoder are specified. The encoder is not specified, to allow for future improvements. All layers work with similar bit-rates:

Layer I:	32–448kbps
Layer II:	8–384kbps
Layer III:	8–320kbps

Two audio channels are supported with four modes of operation:

- normal stereo
- joint (intensity and/or ms) stereo dual
- channel mono
- single channel mono

For normal stereo, one channel carries the left audio signal and one channel carries the right audio signal. For intensity stereo (supported by all layers), high frequencies (above 2 kHz) are combined. The stereo image is preserved but only the temporal envelope is transmitted. For ms stereo (supported by Layer III only), one channel carries the sum signal (L + R) and the other the difference (L–R) signal. In addition, pre-emphasis, copyright marks, and original/copy indication are supported.

FAQs

How do you determine which layer to use for highest sound quality?

To determine which layer should be used for a specific application, look at the available bit-rate, as each layer was designed to support certain bit-rates with a minimum degradation of sound quality.

Layer I, a simplified version of Layer 2, has a target bit-rate 192 kbps per channel or higher.

Layer II is identical to MUSICAM, and has a target bit-rate 128 kbps per channel. It was designed as a trade-off between sound quality and encoder complexity. It is most useful for bit-rates around 96–128 kbps per channel.

Layer III merges the best ideas of MUSICAM and ASPEC and has a target bit-rate of about 64 kbps per channel. The Layer III format specifies a set of advanced features that all address a single goal: to preserve as much sound quality as possible, even at relatively low bit-rates.

Insider Info

Layer III is also known as MP3, the popular music/audio standard.

Video Coding

MPEG-1 permits resolutions up to 4095 × 4095 at 60 frames per second (progressive scan). What many people think of as MPEG-1 is a subset known as Constrained Parameters Bitstream (CPB). The CPB is a limited set of sampling and bit-rate parameters designed to standardize buffer sizes and memory bandwidths, allowing a nominal guarantee of interoperability for decoders and encoders, while still addressing the widest possible range of applications.

TABLE 7.1 Some of the Constrained Parameters for MPEG-1

Horizontal resolution	≤768 samples
Vertical resolution	≤576 scan lines
Picture area	≤396 macroblocks
Pel rate	≤396 × 25 macroblocks per second
Picture rate	≤30 frames per second
Bit-rate	≤1.856 Mbps

TABLE 7.2 Common MPEG-1 Resolutions

Resolution	Frames per Second
352 × 240p	29.97
352 × 240p	23.976
352 × 288p	25
320 × 240p[1]	29.97
384 × 288p[1]	25

Notes: 1. Square pixel format.

Devices not capable of handling these are not considered to be true MPEG-1. Table 7.1 lists some of the constrained parameters. Table 7.2 list some of the more common MPEG-1 resolutions.

Interlaced Video

MPEG-1 was designed to handle progressive (also referred to as noninterlaced) video. Early on, in an effort to improve video quality, several schemes were devised to enable the use of both fields of an interlaced picture.

For example, both fields can be combined into a single frame of 704 × 480p or 704 × 576p resolution and encoded. During decoding, the fields are separated. This, however, results in motion artifacts due to a moving object being in slightly different places in the two fields. Coding the two fields separately avoids motion artifacts, but reduces the compression ratio since the redundancy between fields isn't used.

Encode Preprocessing

Better images can be obtained by preprocessing the video stream prior to MPEG encoding.

To avoid serious artifacts during encoding of a particular picture, prefiltering can be applied over the entire picture or just in specific problem areas. Prefiltering before compression processing is analogous to anti-alias filtering prior to A/D conversion. Prefiltering may take into account texture patterns, motion, and edges, and may be applied at the picture, slice, macroblock, or block level.

MPEG encoding works best on scenes with little fast or random movement and good lighting. For best results, foreground lighting should be clear and background lighting diffused. Foreground contrast and detail should be normal, but low contrast backgrounds containing soft edges are preferred. Editing tools typically allow you to preprocess potential problem areas.

Coded Frame Types

There are four types of coded frames. I (intra) frames (~1 bit/pixel) are frames coded as a stand-alone still image. They allow random access points within the video stream. As such, I frames should occur about two times a second. I frames should also be used where scene cuts occur.

P (predicted) frames (~0.1 bit/pixel) are coded relative to the nearest previous I or P frame, resulting in forward prediction processing, as shown in Figure 7.1. P frames provide more compression than I frames, through the use of motion compensation, and are also a reference for B frames and future P frames.

B (bidirectional) frames (~0.015 bit/pixel) use the closest past and future I or P frame as a reference, resulting in bi-directional prediction, as shown

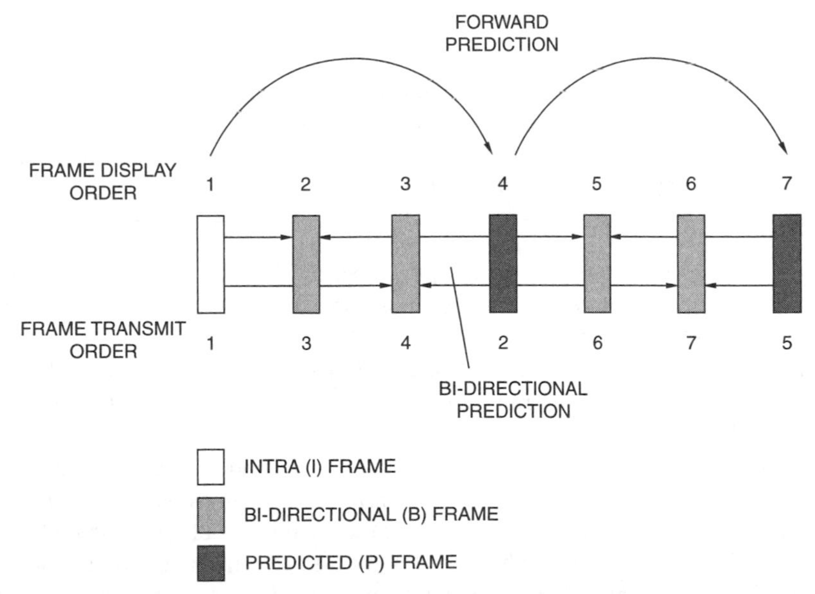

FIGURE 7.1 MPEG-1 I, P, and B Frames. Some frames are transmitted out of display sequence, complicating the interpolation process, and requiring frame reordering by the MPEG decoder. Arrows show inter-frame dependencies.

in Figure 7.1. B frames provide the most compression and decrease noise by averaging two frames. Typically, there are two B frames separating I or P frames.

D (DC) frames are frames coded as a stand-alone still image, using only the DC component of the DCTs. D frames may not be in a sequence containing any other frame types and are rarely used.

A *group of pictures* (GOP) is a series of one or more coded frames intended to assist in random accessing and editing. The GOP value is configurable during the encoding process. The smaller the GOP value, the better the response to movement (since the I frames are closer together), but the lower the compression.

In the coded bitstream, a GOP must start with an I frame and may be followed by any number of I, P, or B frames in any order. In display order, a GOP must start with an I or B frame and end with an I or P frame. Thus, the smallest GOP size is a single I frame, with the largest size unlimited.

Originally, each GOP was to be coded and displayed independently of any other GOP. However, this is not possible unless no B frames precede I frames, or if they do, they use only backward motion compensation. This results in both open and closed GOP formats. A closed GOP is a GOP that can be decoded without using frames of the previous GOP for motion compensation. An open GOP requires that they be available.

Motion Compensation

Key Concept

Motion compensation improves compression of P and B frames by removing temporal redundancies between frames. It works at the macroblock level (a macroblock is 16 samples by 16 lines of y components and the corresponding two 8-sample by 8-line Cb and Cr components).

The motion compensation technique relies on the fact that within a short sequence of the same general image, most objects remain in the same location, while others move only a short distance. The motion is described as a two-dimensional motion vector that specifies where to retrieve a macroblock from a previously decoded frame to predict the sample values of the current macroblock.

After a macroblock has been compressed using motion compensation, it contains both the spatial difference (motion vectors) and content difference (error terms) between the reference macroblock and macroblock being coded.

Note that there are cases where information in a scene cannot be predicted from the previous scene, such as when a door opens. The previous scene doesn't contain the details of the area behind the door. In cases such as this, when a macroblock in a P frame cannot be represented by motion compensation,

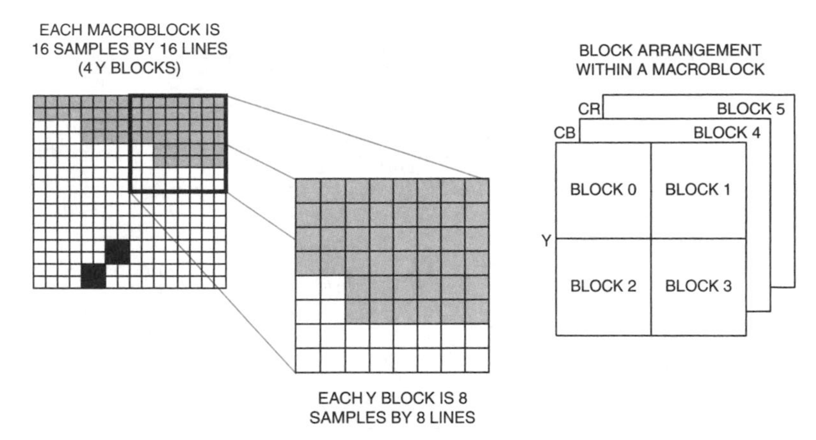

FIGURE 7.2 MPEG-1 Macroblocks and Blocks.

it is coded the same way as a macroblock in an I frame (using intra-picture coding).

Macroblocks in B frames are coded using either the closest previous or future I or P frames as a reference, resulting in four possible codings:

- intra-coding
  - no motion compensation
- forward prediction
  - closest previous I or P frame is the reference
- backward prediction
  - closest future I or P frame is the reference
- bi-directional prediction
  - two frames are used as the reference:
  - the closest previous I or P frame and
  - the closest future I or P frame

Video Bitstream

Figure 7.3 illustrates the video bitstream, a hierarchical structure with seven layers. From top to bottom the layers are:

1. Video Sequence
2. Sequence Header
3. Group of Pictures (GOP)
4. Picture
5. Slice
6. Macroblock (MB)
7. Block

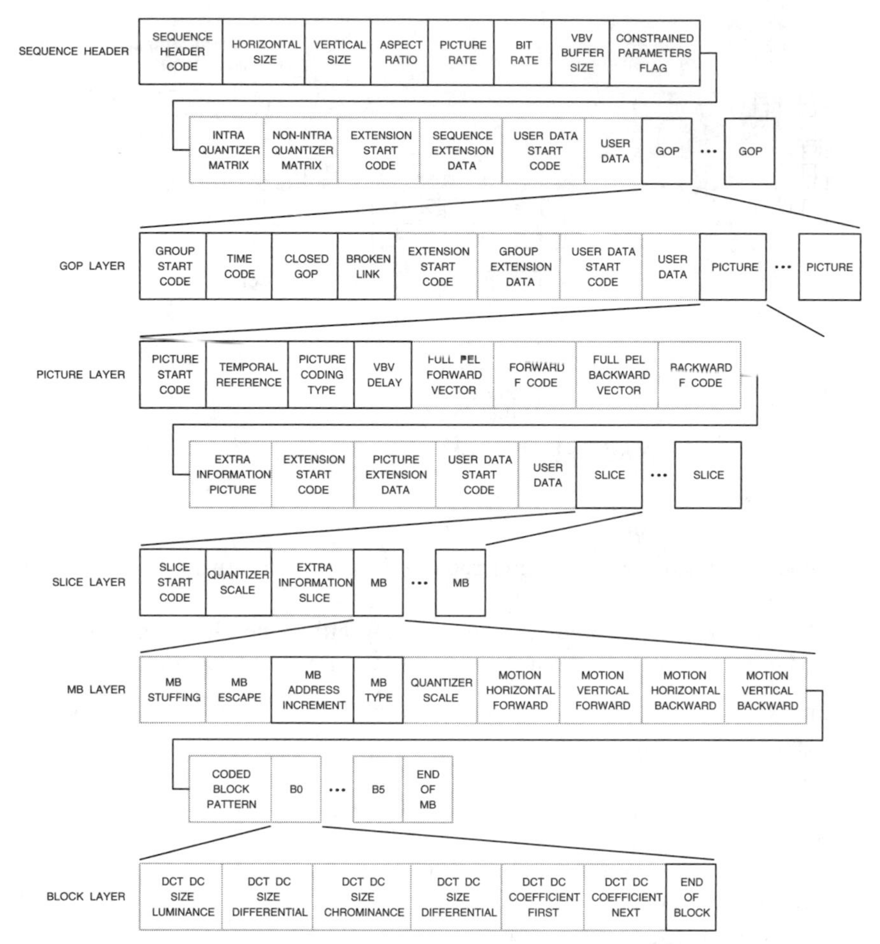

FIGURE 7.3 MPEG-1 Video Bitstream Layer Structures. Market and reserved bits not shown.

Video Decoding

A system demultiplexer parses the system bitstream, demultiplexing the audio and video bitstreams. The video decoder essentially performs the inverse of the encoder. From the coded video bitstream, it reconstructs the I frames. Using I frames, additional coded data, and motion vectors, the P and B frames are generated. Finally, the frames are output in the proper order.

MPEG-2

MPEG-2 extends MPEG-1 to cover a wider range of applications. The primary application targeted during the definition process was all-digital transmission of broadcast-quality video at bit-rates of 4–9 Mbps. However, MPEG-2 is useful for many other applications, such as HDTV, and now supports bit-rates of 1.5–60 Mbps.

MPEG-2 is an ISO standard (ISO/IEC 13818), and consists of eleven parts:

Systems	ISO/IEC 13818–1
Video	ISO/IEC 13818–2
Audio	ISO/IEC 13818–3
Conformance testing	ISO/IEC 13818–4
Software simulation	ISO/IEC 13818–5
DSM-CC extensions	ISO/IEC 13818–6
Advanced audio coding	ISO/IEC 13818–7
RTI extension	ISO/IEC 13818–9
DSM-CC conformance	ISO/IEC 13818–10
IPMP	ISO/IEC 13818–11

As with MPEG-1, the compressed bitstreams implicitly define the decompression algorithms. The compression algorithms are up to the individual manufacturers, within the scope of an international standard.

The Digital Storage Media Command and Control (DSM-CC) extension (ISO/IEC 13818–6) is a toolkit for developing control channels associated with MPEG-2 streams. In addition to providing VCR-type features such as fast-forward, rewind, pause, etc., it may be used for a wide variety of other purposes, such as packet data transport. DSM-CC works in conjunction with next-generation packet networks, working alongside Internet protocols as RSVP, RTSP, RTP, and SCP.

The Real Time Interface (RTI) extension (ISO/IEC 13818-9) defines a common interface point to which terminal equipment manufacturers and network operators can design. RTI specifies a delivery model for the bytes of an MPEG-2 System stream at the input of a real decoder, whereas MPEG-2 System defines an idealized byte delivery schedule.

IPMP (Intellectual Property Management and Protection) is a digital rights management (DRM) standard, adapted from the MPEG-4 IPMP extension specification. Rather than a complete system, a variety of functions are provided within a framework.

Audio Overview

In addition to the non-backwards-compatible audio extension (ISO/IEC 13818–7), MPEG-2 supports up to five full-bandwidth channels compatible with MPEG-1 audio coding. It also extends the coding of MPEG-1 audio to half sampling rates (16 kHz, 22.05 kHz, and 24 kHz) for improved quality for bit-rates at or below 64 kbps per channel.

Insider Info

MPEG-2.5 is an unofficial, yet common, extension to the audio capabilities of MPEG-2. It adds sampling rates of 8 kHz, 11.025 kHz, and 12 kHz.

Video Overview

With MPEG-2, profiles specify the syntax (i.e., algorithms) and levels specify various parameters (resolution, frame rate, bit-rate, etc.). Main Profile@Main Level is targeted for SDTV applications, while Main Profile@High Level is targeted for HDTV applications.

Levels

MPEG-2 supports four levels, which specify resolution, frame rate, coded bit-rate, and so on for a given profile.

Low Level (LL)

MPEG-1 Constrained Parameters Bit-stream (CPB), supporting up to 352 × 288 at up to 30 frames per second. Maximum bit-rate is 4 Mbps.

Main Level (ML)

MPEG-2 Constrained Parameters Bit-stream (CPB) supports up to 720 × 576 at up to 30 frames per second and is intended for SDTV applications. Maximum bit-rate is 15–20 Mbps.

High 1440 Level

This level supports up to 1440 × 1088 at up to 60 frames per second and is intended for HDTV applications. Maximum bit-rate is 60–80 Mbps.

High Level (HL)

High Level supports up to 1920 × 1088 at up to 60 frames per second and is intended for HDTV applications. Maximum bit-rate is 80–100 Mbps.

Profiles

MPEG-2 supports six profiles, which specify which coding syntax (algorithms) is used. Tables 7.3 through 7.10 illustrate the various combinations of levels and profiles allowed.

Simple Profile (SP)

Main profile without the B frames, intended for software applications and perhaps digital cable TV.

Main Profile (MP)

Supported by most MPEG-2 decoder chips, it should satisfy 90% of the consumer SDTV and HDTV applications. Typical resolutions are shown in Table 7.8.

Multiview Profile (MVP)

By using existing MPEG-2 tools, it is possible to encode video from two cameras shooting the same scene with a small angle difference.

4:2:2 Profile (422P)

Previously known as "studio profile," this profile uses 4:2:2 YCbCr instead of 4:2:0, and with main level, increases the maximum bit-rate up to 50 Mbps (300 Mbps with high level). It was added to support pro-video SDTV and HDTV requirements.

SNR and Spatial Profiles

Adds support for SNR scalability and/or spatial scalability.

High Profile (HP)

Targeted for pro-video HDTV applications.

Scalability

The MPEG-2 SNR, Spatial, and High profiles support four scalable modes of operation. These modes break MPEG-2 video into layers for the purpose of prioritizing video data.

TABLE 7.3 MPEG-2 Acceptable Combinations of Levels and Profiles

Level	Profile						
	Nonscalable				Scalable		
	Simple	Main	Multiview	4:2:2	SNR	Spatial	High
High	–	yes	–	yes	–	–	yes
High 1440	–	yes	–	–	–	yes	yes
Main	yes	yes	yes	yes	yes	–	yes
Low	–	yes	–	–	yes	–	–

TABLE 7.4 Some MPEG-2 Profile Constraints

Constraint	Profile						
	Nonscalable				Scalable		
	Simple	Main	Multiview	4:2:2	SNR	Spatial	High
Chroma format	4:2:0	4:2:0	4:2:0	4:2:0 or 4:2:2	4:2:0	4:2:0	4:2:0 or 4:2:2
Picture types	I, P	I, P, B	I, P, B	I, P, B	I, P, B	I, P, B	I, P, B
Scalable modes	–	–	Temporal	–	SNR	SNR or Spatial	SNR or Spatial
Intra dc precision (bits)	8, 9, 10	8, 9, 10	8, 9, 10	8,9,10,11	8,9,10	8,9,10	8,9,10,11
Sequence scalable extension	no	no	yes	no	yes	yes	yes
Picture spatial scalable extension	no	no	no	no	no	yes	yes
Picture temporal scalable extension	no	no	yes	no	no	no	no
Repeat first field	constrained		unconstrained	constrained	unconstrained		

TABLE 7.5 MPEG-2 Number of Permissible Layers for Scalable Profiles

Level	Maximum Number of Layers	Profile			
		SNR	Spatial	High	Multiview
High	All layers (base + enhancement)	–	–	3	2
	Spatial enhancement layers			1	0
	SNR enhancement layers			1	0
	Temporal auxiliary layers			0	1
High 1440	All layers (base + enhancement)	–	3	3	2
	Spatial enhancement layers		1	1	0
	SNR enhancement layers		1	1	0
	Temporal auxiliary layers		0	0	1
Main	All layers (base + enhancement)	2	–	3	2
	Spatial enhancement layers	0		1	0
	SNR enhancement layers	1		1	0
	Temporal auxiliary layers	0		0	1
Low	All layers (base + enhancement)	2	–	–	2
	Spatial enhancement layers	0			0
	SNR enhancement layers	1			0
	Temporal auxiliary layers	0			1

ALERT!

Scalability is not commonly used since efficiency decreases by about 2 dB (or about 30% more bits are required).

SNR Scalability

This mode is targeted for applications that desire multiple quality levels. All layers have the same spatial resolution. The base layer provides the basic video quality. The enhancement layer increases the video quality by providing refinement data for the DCT coefficients of the base layer.

Spatial Scalability

Useful for simulcasting, each layer has a different spatial resolution. The base layer provides the basic spatial resolution and temporal rate. The enhancement layer uses the spatially interpolated base layer to increase the spatial resolution. For example, the base layer may implement 352 × 240 resolution video, with the enhancement layers used to generate 704 × 480 resolution video.

TABLE 7.6 Some MPEG-2 Video Decoder Requirements for Various Profiles

Profile	Profile			Profile at Level for Base Decoder
	Base Layer	Enhancement Layer 1	Enhancement Layer 2	
SNR	4:2:0	SNR, 4:2:0	–	MP@same level
Spatial	4:2:0	SNR, 4:2:0	–	MP@same level
	4:2:0	Spatial, 4:2:0	–	MP@(level–1)
	4:2:0	SNR, 4:2:0	Spatial, 4:2:0	
	4:2:0	Spatial, 4:2:0	SNR, 4:2:0	
High	4:2:0 or 4:2:2	–	–	MP@same level
	4:2:0	SNR, 4:2:0	–	
	4:2:0 or 4:2:2	SNR, 4:2:2	–	HP@ (level–1)
	4:2:0	Spatial, 4:2:0	–	
	4:2:0 or 4:2:2	Spatial, 4:2:2	–	
	4:2:0	SNR, 4:2:0	Spatial, 4:2:0 or 4:2:2	
	4:2:0 or 4:2:2	SNR, 4:2:2	Spatial, 4:2:2	
	4:2:0	Spatial, 4:2:0	SNR, 4:2:0 or 4:2:2	
	4:2:0	Spatial, 4:2:2	SNR, 4:2:2	
	4:2:2	Spatial, 4:2:2	SNR, 4:2:2	
Multiview	4:2:0	Temporal, 4:2:0	–	MP@same level

Temporal Scalability

This mode allows migration from low temporal rate to higher temporal rate systems. The base layer provides the basic temporal rate. The enhancement layer uses temporal prediction relative to the base layer. The base and enhancement layers can be combined to produce a full temporal rate output. All layers have the same spatial resolution and chroma formats. In case of errors in the enhancement layers, the base layer can be used for concealment.

Data Partitioning

This mode is targeted for cell loss resilience in ATM networks. It breaks the 64 quantized transform coefficients into two bitstreams. The higher priority bitstream contains critical lower-frequency DCT coefficients and side information such as headers and motion vectors. A lower-priority bitstream carries higher-frequency DCT coefficients that add detail.

TABLE 7.7 **MPEG-2 Upper Limits of Resolution and Temporal Parameters. In the case of single layer or SNR scalability coding, the "Enhancement Layer" parameters apply**

Level	Spatial Resolution Layer	Parameter	Profile					
			Simple	Main	Multiview	4:2:2	SNR / Spatial	High
High	Enhancement	Samples per line	–	1920	1920	1920	–	1920
		Lines per frame		1088	1088	1088		1088
		Frames per second		60	60	60		60
	Lower	Samples per line	–	–	1920	–	–	960
		Lines per frame			1088			576
		Frames per second			60			30
High 1440	Enhancement	Samples per line	–	1440	1440	–	1440	1440
		Lines per frame		1088	1088		1088	1088
		Frames per second		60	60		60	60
	Lower	Samples per line	–	–	1440	–	720	720
		Lines per frame			1088		576	576
		Frames per second			60		30	30
Main	Enhancement	Samples per line	720	720	720	720	720	720
		Lines per frame	576	576	576	608	576	576
		Frames per second	30	30	30	30	30	30
	Lower	Samples per line	–	–	720	–	–	352
		Lines per frame			576			288
		Frames per second			30			30
Low	Enhancement	Samples per line	–	352	352	–	352	–
		Lines per frame		288	288		288	
		Frames per second		30	30		30	
	Lower	Samples per line	–	–	352	–	–	–
		Lines per frame			288			
		Frames per second			30			

Note: 1. The above levels and profiles that originally specified 1152 maximum lines per frame were changed to 1088 lines per frame.

TABLE 7.8 Example Levels and Resolutions for MPEG-2 Main Profile

Level	Maximum Bit-Rate (Mbps)	Typical Active Resolutions	Frame Rate (Hz)[2]										
			23.976p	24p	25p	29.97p	30p	50p	59.94p	60p	25i	29.97i	30i
High	80 (100 for High Profile) (300 for 4:2:2 Profile)	1920 × 1080[1]	×	×	×	×	×				×	×	×
High 1440	60 (80 for High Profile)	1280 × 720	×	×	×	×	×	×	×	×			
		960 × 1080[1]	×	×	×	×	×				×	×	×
		1280 × 1080[1]	×	×	×	×	×				×	×	×
		1440 × 1080[1]	×	×	×	×	×				×	×	×
Main	15 (20 for High Profile) (50 for 4:2:2 Profile)	352 × 480	×	×		×	×		×	×		×	×
		352 × 576		×	×			×			×		
		480 × 480	×	×		×	×		×	×		×	×
		544 × 480	×	×		×	×		×	×		×	×
		544 × 576		×	×			×			×		
		640 × 480	×	×		×	×		×	×		×	×
		704 × 480, 720 × 480	×	×		×	×		×	×		×	×
		704 × 576, 720 × 576		×	×			×			×		
Low	4	320 × 240	×	×		×	×		×	×		×	×
		352 × 240	×	×		×	×		×	×		×	×
		352 × 288		×	×			×			×		

Notes: 1. The video coding system requires that the number of active scan lines be a multiple of 32 for interlaced pictures, and a multiple of 16 for progressive pictures. Thus, for the 1080-line inter-laced format, the video encoder and decoder must actually use 1088 lines. The extra eight lines are "dummy" lines having no content, and designers choose dummy data that simplifies the implementation. The extra eight lines are always the last eight lines of the encoded image. These dummy lines do not carry useful information, but add little to the data required for transmission.
2. p = progressive; i = interlaced.

TABLE 7.9 **MPEG-2 Upper Limits for Y Sample Rate (M samples/second). In the case of single layer or SNR scalability coding, the "Enhancement Layer" parameters apply**

Level	Spatial Resolution Layer	Profile					
		Simple	Main	Multiview	SNR / Spatial	High	4:2:2
High	Enhancement	–	62.668800	62.668800	–	62.668800 (4:2:2) 83.558400 (4:2:0)	62.668800
	Lower	–	–	62.668800	–	14.745600 (4:2:2) 19.660800 (4:2:0)	–
High 1440	Enhancement	–	47.001600	47.001600	47.001600	47.001600 (4:2:2) 62.668800 (4:2:0)	–
	Lower	–	–	47.001600	10.368000	11.059200 (4:2:2) 14.745600 (4:2:0)	–
Main	Enhancement	10.368000	10.368000	10.368000	10.368000	11.059200 (4:2:2) 14.745600 (4:2:0)	11.059200
	Lower	–	–	10.368000	–	3.041280 (4:2:0)	–
Low	Enhancement	–	3.041280	3.041280	3.041280	–	–
	Lower	–	–	3.041280	–	–	–

TABLE 7.10 MPEG-2 Upper Limits for Bit-Rates (Mbps)

Level	Profile					
	Nonscalable				Scalable	
	Simple	Main	Multiview	4:2:2	SNR/Spatial	High
High	–	80	130 (both layers) 80 (base layer)	300	–	100 (all layers) 80 (middle + base layers) 25 (base layer)
High 1440	–	60	100 (both layers) 60 (base layer)	–	60 (all layers) 40 (middle + base layers) 15 (base layer)	80 (all layers) 60 (middle + base layers) 20 (base layer)
Main	15	15	25 (both layers) 5 (base layer)	50	15 (both layers) 10 (base layer)	20 (all layers) 15 (middle + base layers) 4 (base layer)
Low	–	4	8 (both layers) 4 (base layer)	–	4 (both layers) 3 (base layer)	–

Transport and Program Streams

The MPEG-2 Systems Standard specifies two methods for multiplexing the audio, video, and other data into a format suitable for transmission and storage.

The program stream is designed for applications where errors are unlikely. It contains audio, video, and data bitstreams (also called elementary bitstreams) all merged into a single bitstream. The program stream, as well as each of the elementary bitstreams, may be a fixed or variable bit-rate. DVDs and SVCDs use program streams, carrying the DVD- and SVCD-specific data in private data streams interleaved with the video and audio streams.

The transport stream, using fixed-size packets of 188 bytes, is designed for applications where data loss is likely. Also containing audio, video, and data bitstreams all merged into a single bitstream, multiple programs can be carried. The ARIB, ATSC, DVB and Open-Cable™ standards use transport streams.

Both the transport stream and program stream are based on a common packet structure, facilitating common decoder implementations and conversions. Both streams are designed to support a large number of known and anticipated applications, while retaining flexibility.

Video Coding Layer

YCbCr Color Space

MPEG-2 uses the YCbCr color space, supporting 4:2:0, 4:2:2, and 4:4:4 sampling. The 4:2:2 and 4:4:4 sampling options increase the chroma resolution over 4:2:0, resulting in better picture quality.

Coded Picture Types

There are three types of coded pictures. *I* (*intra*) *pictures* are fields or frames coded as a stand-alone still image. They allow random access points within the video stream. As such, I pictures should occur about two times a second. I pictures also should be used where scene cuts occur.

P (*predicted*) *pictures* are fields or frames coded relative to the nearest previous I or P picture, resulting in forward prediction processing. P pictures provide more compression than I pictures, through the use of motion compensation, and are also a reference for B pictures and future P pictures.

B (*bidirectional*) *pictures* are fields or frames that use the closest past and future I or P picture as a reference, resulting in bidirectional prediction, as shown in Figure 7.1. B pictures provide the most compression and decrease noise by averaging two pictures. Typically, there are two B pictures separating I or P pictures.

D (*DC*) pictures are not supported in MPEG-2, except for decoding to support backwards compatibility with MPEG-1.

A *group of pictures* (GOP) is a series of one or more coded pictures intended to assist in random accessing and editing. The GOP value is configurable during the encoding process.

ALERT!

The smaller the GOP value, the better the response to movement (since the I pictures are closer together), but the lower the compression.

In the coded bitstream, a GOP must start with an I picture and may be followed by any number of I, P, or B pictures in any order. In display order, a GOP must start with an I or B picture and end with an I or P picture. Thus, the smallest GOP size is a single I picture, with the largest size unlimited.

Each GOP should be coded independently of any other GOP. However, this is not true unless no B pictures precede the first I picture, or if they do, they use only backward motion compensation. This results in both open and closed GOP formats. A closed GOP is a GOP that can be decoded without using pictures of the previous GOP for motion compensation. An open GOP, identified by the broken_link flag, indicates that the first B pictures (if any) immediately following the first I picture after the GOP header may not be decoded correctly (and thus not be displayed) since the reference picture used for prediction is not available due to editing.

Motion Compensation

Motion compensation for MPEG-2 is more complex due to the introduction of fields. After a macroblock has been compressed using motion compensation, it contains both the spatial difference (motion vectors) and content difference (error terms) between the reference macroblock and macroblock being coded.

The two major classifications of prediction are field and frame. Within field pictures, only field predictions are used. Within frame pictures, either field or frame predictions can be used (selectable at the macroblock level).

Motion vectors for MPEG-2 are always coded in half-pixel units. MPEG-1 supports either half-pixel or full-pixel units.

16 × 8 Motion Compensation Option

Two motion vectors (four for B pictures) per macroblock are used, one for the upper 16 × 8 region of a macroblock and one for the lower 16 × 8 region of a macroblock. It is only used with field pictures.

Dual-Prime Motion Compensation Option

This is only used with P pictures that have no B pictures between the predicted and reference fields of frames. One motion vector is used, together with a small differential motion vector. All of the necessary predictions are derived from these.

Macroblocks

Three types of macroblocks are available in MPEG-2.

The 4:2:0 macroblock consists of four Y blocks, one Cb block, and one Cr block. The block ordering is shown in the figure.

The 4:2:2 macroblock consists of four Y blocks, two Cb blocks, and two Cr blocks. The block ordering is shown in the figure.

The 4:4:4 macroblock consists of four Y blocks, four Cb blocks, and four Cr blocks. The block ordering is shown in the figure.

Macroblocks in P pictures are coded using the closest previous I or P picture as a reference, resulting in two possible codings:

- intra-coding
  - no motion compensation
- forward prediction
  - closest previous I or P picture is the reference

Macroblocks in B pictures are coded using the closest previous and/or future I or P picture as a reference, resulting in four possible codings:

- intra-coding
  - no motion compensation
- forward prediction
  - closest previous I or P picture is the reference
- backward prediction
  - closest future I or P picture is the reference
- bidirectional prediction
  - two pictures used as the reference:
  - the closest previous I or P picture and
  - the closest future I or P picture

Decoder Considerations

The video decoder essentially performs the inverse function of the encoder. From the coded bitstream, it reconstructs the I frames. Using I frames, additional coded data, and motion vectors, the P and B frames are generated. Finally, the frames are output in the proper order.

Figure 7.4 illustrates the block diagram of a basic MPEG-2 video decoder.

MPEG-4

MPEG-4 builds upon the success and experience of MPEG-2. It is best known for:

- Lower bit-rates than MPEG-2 (for the same quality of video)
- Use of natural or synthetic objects that can be rendered together to make a scene
- Support for interactivity

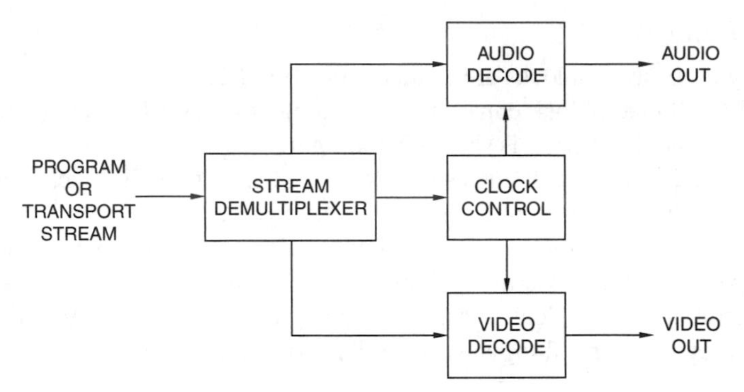

FIGURE 7.4 Simplified MPEG-2 Decoder Block Diagram.

For authors, MPEG-4 enables creating content that is more reusable and flexible, with better content protection capabilities.

For consumers, MPEG-4 can offer more interactivity and, due to the lower bit-rate over MPEG-2, the ability to enjoy content over new networks (such as DSL) and mobile products.

MPEG-4 is an ISO standard (ISO/IEC 14496).

MPEG-4 provides a standardized way to represent audio, video, or still image media objects using descriptive elements (instead of actual bits of an image, for example). A media object can be natural or synthetic (computer-generated) and can be represented independent of its surroundings or background.

It also describes how to merge multiple media objects to create a scene. Rather than sending bits of picture, the media objects are sent, and the receiver composes the picture. This allows:

- An object to be placed anywhere
- Geometric transformations on an object
- Grouping of objects
- Modifying attributes and transform data
- Changing the view of a scene dynamically

Audio Overview

MPEG-4 audio supports a wide variety of applications, from simple speech to multi-channel high-quality audio.

Audio objects (audio codecs) use specific combinations of tools to efficiently represent different types of audio objects. Profiles use specific combinations of audio object types to efficiently service a specific market segment. Levels specify size, rate, and complexity limitations within a profile to ensure interoperability.

Currently, most solutions support a few of the most popular audio codecs (usually AAC-LC and HE-AAC) rather than one or more profiles/levels.

Visual Overview

MPEG-4 visual is divided into two sections. MPEG-4.2 includes the original MPEG-4 video codecs discussed in this section. MPEG-4.10 specifies the "advanced video codec," also known as H.264, and is discussed in the next section.

The visual specifications are optimized for three primary bit-rate ranges:

- less than 64 kbps
- 64–384 kbps
- 0.384–4 Mbps

For high-quality applications, higher bit-rates are possible, using the same tools and bitstream syntax as those used for lower bit-rates.

With MPEG-4, visual objects (video codecs) use specific combinations of tools to efficiently represent different types of visual objects. Profiles use specific combinations of visual object types to efficiently service a specific market segment. Levels specify size, rate, and complexity limitations within a profile to ensure interoperability.

Insider Info

Currently, most solutions support only a couple of the MPEG-4.2 video codecs (usually Simple and Advanced Simple) due to silicon cost issues. Interest in MPEG-4.2 video codecs also dropped dramatically with the introduction of the MPEG-4.10 (H.264) and SMPTE 421M (VC-1) video codecs, which offer about two times better performance.

YCbCr Color Space

The 4:2:0 YCbCr color space is used for most objects. Each component can be represented by a number of bits ranging from 4 to 12 bits, with 8 bits being the most commonly used.

MPEG-4.2 Simple Studio and Core Studio objects may use 4:2:2, 4:4:4, 4:2:2:4, and 4:4:4:4:4:4 YCbCr or RGB sampling options, to support the higher picture quality required during the editing process.

Like H.263 and MPEG-2, the MPEG-4.2 video codecs are also macroblock, block, and DCT-based.

Visual Objects

Instead of the video frames or pictures used in earlier MPEG specifications, MPEG-4 uses natural and synthetic visual objects. Instances of video objects at a given time are called *visual object planes* (VOPs).

Much like MPEG-2, there are I (intra), P (predicted), and B (bidirectional) VOPs. The S-VOP is a VOP for a sprite object. The S(GMC)-VOP is coded using prediction based on global motion compensation from a past reference VOP.

Arbitrarily shaped video objects, as well as rectangular objects, may be used. An MPEG-2 video stream can be a rectangular video object, for example.

Objects may also be scalable, enabling the reconstruction of useful video from pieces of a total bitstream. This is done by using a base layer and one or more enhancement layers.

Only natural visual object types are discussed since they are currently of the most interest in the marketplace.

MPEG-4.2 Natural Visual Object Types

MPEG-4.2 supports many natural visual object types (video codecs), with several interesting ones shown in Table 7.11. The more common object types are:

Main Objects

Main objects provide the highest video quality. Compared to Core objects, they also support grayscale shapes, sprites, and both interlaced and progressive content.

Core Objects

Core objects use a subset of the tools used by Main objects, although B-VOPs are still supported. They also support scalability by sending extra P-VOPs. Binary shapes can include a constant transparency but cannot do the variable transparency offered by grayscale shape coding.

Simple Objects

Simple objects are low bit-rate, error resilient, rectangular natural video objects of arbitrary aspect ratio. Simple objects use a subset of the tools used by Core objects.

Advanced Simple Objects

Advanced Simple objects looks much like Simple objects in that only rectangular objects are supported, but adds a few tools to make it more efficient: B-frames, ¼-pixel motion compensation (QPEL), and global motion compensation (GMC).

Fine Granularity Scalable Objects

Fine Granularity Scalable objects can use up to eight scalable layers so delivery quality can easily adapt to transmission and decoding circumstances.

TABLE 7.11 Available Tools for Common MPEG-4.2 Natural Visual Object Types

Tools	Object Type						
	Main	Core	Simple	Advanced Simple	Advanced Real Time Simple	Advanced Coding Efficiency	Fine Granularity Scalable
VOP types	I, P, B	I, P, B	I, P	I, P, B	I, P	I, P, B	I, P, B
Chroma format	4:2:0	4:2:0	4:2:0	4:2:0	4:2:0	4:2:0	4:2:0
Interlace	×	–	–	×	–	×	×
Global motion compensation (GMC)	–	–	–	×	–	×	–
Quarter-pel motion compensation (QPEL)	–	–	–	×	–	×	–
Slice resynchronization	×	×	×	×	×	×	×
Data partitioning	×	×	×	×	×	×	×
Reversible VLC	×	×	×	×	×	×	×
Short header	×	×	×	×	×	×	×
Method 1 and 2 quantization	×	×	–	×	–	×	×
Shape adaptive DCT	–	–	–	–	–	×	–
Dynamic resolution conversion	–	–	–	–	×	×	–
NEWPRED	–	–	–	–	×	×	–
Binary shape	×	×	–	–	–	×	–
Grey shape	×	–	–	–	–	×	–
Sprite	×						
Fine granularity scalability (FGS)	–	–	–	–	–	–	×
FGS temporal scalability	–	–	–	–	–	–	×

Graphics Overview

Graphics profiles specify which graphics elements of the BIFS tool can be used to build a scene. Although it is defined in the Systems specification, graphics is really just another media profile like audio and video, so it is discussed here.

Four hierarchical graphics profiles are defined: Simple 2D, Complete 2D, Complete and 3D Audio Graphics. They differ in the graphics elements of the BIFS tool to be supported by the decoder.

Simple 2D profile provides the basic features needed to place one or more visual objects in a scene.

Complete 2D profile provides 2D graphics functions and supports features such as arbitrary 2D graphics and text, possibly in conjunction with visual objects.

Complete profile provides advanced capabilities such as elevation grids, extrusions, and sophisticated lighting. It enables complex virtual worlds to exhibit a high degree of realism.

3D Audio Graphics profile may be used to define the acoustical properties of the scene (geometry, acoustics absorption, diffusion, material transparency). This profile is useful for applications that do environmental equalization of the audio signals.

Visual Layers

An MPEG-4 visual scene consists of one or more video objects. Currently, the most common video object is a simple rectangular frame of video.

Each video object may have one or more layers to support temporal or spatial scalable coding. This enables the reconstruction of video in a layered manner, starting with a base layer and adding a number of enhancement layers. Where a high degree of scalability is needed, such as when an image is mapped onto a 2D or 3D object, a wavelet transform is available.

The visual bitstream provides a hierarchical description of the scene. Each level of hierarchy can be accessed through the use of unique start codes in the bitstream.

Visual Object Sequence (VS)

This is the complete scene which contains all the 2D or 3D, natural or synthetic, objects and any enhancement layers.

Video Object (VO)

A video object corresponds to a particular object in the scene. In the most simple case, this can be a rectangular frame, or it can be an arbitrarily shaped object corresponding to an object or background of the scene.

Video Object Layer (VOL)

Each video object can be encoded in scalable (multi-layer) or nonscalable form (single layer), depending on the application, represented by the video object layer (VOL). The VOL provides support for scalable coding. A video object can be encoded using spatial or temporal scalability, going from coarse to fine resolution. Depending on parameters such as available bandwidth, computational power, and user preferences, the desired resolution can be made available to the decoder.

There are two types of video object layers, the video object layer that provides full MPEG-4 functionality, and a reduced functionality video object layer, the video object layer with short headers. The latter provides bitstream compatibility with baseline H.263.

Group of Video Object Plane (GOV)

Each video object is sampled in time; each time sample of a video object is a video object plane. Video object planes can be grouped together to form a group of video object planes.

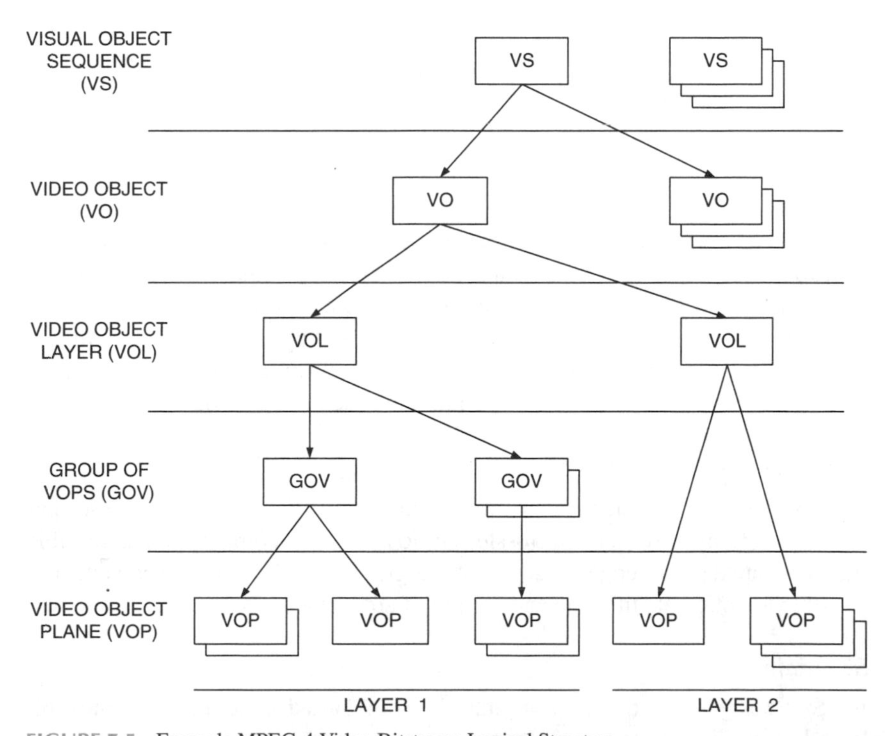

FIGURE 7.5 Example MPEG-4 Video Bitstream Logical Structure.

The GOV groups together video object planes. GOVs can provide points in the bitstream where video object planes are encoded independently from each other, and can thus provide random access points into the bitstream. GOVs are optional.

Video Object Plane (VOP)

A VOP is a time sample of a video object. VOPs can be encoded independently of each other, or dependent on each other by using motion compensation. A conventional video frame can be represented by a VOP with rectangular shape.

How It Works

Scene Description:

To assemble a multimedia scene at the receiver, it is not sufficient to simply send just the multiple streams of data. For example, objects may be located in 2D or 3D space, and each has its local coordinate system. Objects are positioned within a scene by transforming each of them to the scene's coordinate system. Therefore, additional data is required for the receiver to assemble a meaningful scene for the user. This additional data is called scene description.

Scene graph elements (which are BIFS tools) describe audiovisual primitives and attributes. These elements, and any relationship between them, form a hierarchical scene graph. The scene graph is not necessarily static; elements may be added, deleted, or modified as needed. The scene graph profile defines the allowable set of scene graph elements that may be used.

BIFS

BIFS (BInary Format for Scenes) is used to not only describe the scene composition information, but also graphical elements. A fundamental difference between the BIFS and VRML is that BIFS is a binary format, whereas VRML is a textual format. BIFS supports the elements used by VRML and several that VRML does not, including compressed binary format, streaming, streamed animation, 2D primitives, enhanced audio, and facial animation.

Compressed Binary Format

BIFS supports an efficient binary representation of the scene graph information. The coding may be either lossless or lossy. Lossy compression is possible due to context knowledge: if some scene graph data has been received, it is possible to anticipate the type and format of subsequent data.

Streaming

BIFS is designed so that a scene may be transmitted as an initial scene, followed by modifications to the scene.

Streamed Animation

BIFS includes a low-overhead method for the continuous animation of changes to numerical values of the elements in a scene. This provides an alternative to the interpolator elements supported in BIFS and VRML.

2D Primitives

BIFS has native support for 2D scenes to support low-complexity, low-cost solutions such as traditional television. Rather than partitioning the world into 2D vs. 3D, BIFS allows both 2D and 3D elements in a single scene.

Enhanced Audio

BIFS improves audio support through the use of an audio scene graph, enabling audio sources to be mixed or the generation of sound effects.

Facial Animation

BIFS exposes the animated face properties to the scene level. This enables it to be a full member of a scene that can be integrated with any other BIFS functionality, similar to other audiovisual objects.

MPEG-4.10 (H.264) VIDEO

Previously known as "H.26L," "JVT," "JVT codec," "AVC," and "Advanced Video codec," ITU-T H.264 is one of two new video codecs, the other being SMPTE 421M (VC-1), which is based on Microsoft Windows Media Video 9 codec. H.264 is incorporated into the MPEG-4 specifications as Part 10.

Rather than a single major advancement, H.264 employs many new tools designed to improve performance. These include:

- Support for 8-, 10-, and 12-bit 4:2:2 and 4:4:4 YCbCr
- Integer transform
- UVLC, CAVLC, and CABAC entropy coding
- Multiple reference frames
- Intra prediction
- In-loop deblocking filter
- SP and SI slices
- Many new error resilience tools

Profiles and Levels

Similar to other video codecs, profiles specify the syntax (i.e., algorithms) and levels specify various parameters (resolution, frame rate, bit-rate, etc.). The various levels are described in Table 7.12.

TABLE 7.12 MPEG-4.10 (H.264) Levels. "MB" = macroblock, "MV" = motion vector

Level	Maximum MB per Second	Maximum Frame Size (MB)	Typical Frame Resolution	Typical Frames per Second	Maximum MVs per Two Consecutive MBs	Maximum Reference Frames	Maximum Bit-Rate
1	1,485	99	176 × 144	15	–	4	64 kbps
1.1	3,000	396	176 × 144 320 × 240 352 × 288	30 10 7.5	–	9 3 3	192 kbps
1.2	6,000	396	352 × 288	15	–	6	384 kbps
1.3	11,880	396	352 × 288	30	–	6	768 kbps
2	11,880	396	352 × 288	30	–	6	2 Mbps
2.1	19,800	792	352 × 480 352 × 576	30 25	–	6	4 Mbps
2.2	20,250	1,620	720 × 480 720 × 576	15 12.5	–	5	4 Mbps
3	40,500	1,620	720 × 480 720 × 576	30 25	32	5	10 Mbps
3.1	108,000	3,600	1280 × 720	30	16	5	14 Mbps
3.2	216,000	5,120	1280 × 720	60	16	4	20 Mbps
4	245,760	8,192	1920 × 1080 1280 × 720	30 60	16	4	20 Mbps
4.1	245,760	8,192	1920 × 1080 1280 × 720	30 60	16	4	50 Mbps
4.2	491,520	8,192	1920 × 1080	60	16	4	50 Mbps
5	589,824	22,080	2048 × 1024	72	16	5	135 Mbps
5.1	983,040	36,864	2048 × 1024 4096 × 2048	120 30	16	5	240 Mbps

Baseline Profile (BP)

Baseline profile is designed for progressive video such as video conferencing, video-over-IP, and mobile applications. Tools used by Baseline profile include:

- I and P slice types
- ¼-pixel motion compensation
- UVLC and CAVLC entropy coding
- Arbitrary slice ordering (ASO)
- Flexible macroblock ordering (FMO)
- Redundant slices (RS)
- 4:2:0 YCbCr format

Insider Info

Note that Baseline profile is not a subset of Main profile. Many solutions implement a subset of Baseline profile, without ASO or FMO; this is a subset of Main profile (and much easier to implement).

Extended Profile (XP)

Extended profile is designed for mobile and Internet streaming applications. Additional tools over Baseline profile include:

- B, SP, and SI slice types
- Slice data partitioning
- Weighted prediction

Main Profile (MP)

Main profile is designed for a wide range of broadcast applications. Additional tools over Baseline profile include:

- Interlaced coding
- B slice type
- CABAC entropy coding
- Weighted prediction
- 4:2:2 and 4:4:4 YCbCr, 10- and 12-bit formats
- ASO, FMO, and RS are not supported

High Profiles (HP)

After the initial specification was completed, the Fidelity Range Extension (FRExt) amendment was added. This resulted in four additional profiles being added to the specification:

- High Profile (HP): adds support for adaptive selection between 4 × 4 and 8 × 8 block sizes for the luma spatial transform and encoder-specified frequency-dependent scaling matrices for transform coefficients

- High 10 Profile (Hi10P): adds support for 9- or 10-bit 4:2:0 YCbCr
- High 4:2:2 Profile (Hi422P): adds support for 4:2:2 YCbCr
- High 4:4:4 Profile (Hi444P): adds support for 11- or 12-bit samples, 4:4:4 YCbCr or RGB, residual color transform and predictive lossless coding

Video Coding Layer

YCbCr Color Space

H.264 uses the YCbCr color space, supporting 4:2:0, 4:2:2, and 4:4:4 sampling. The 4:2:2 and 4:4:4 sampling options increase the chroma resolution over 4:2:0, resulting in better picture quality. In addition to 8-bit YCbCr data, H.264 supports 10- and 12-bit YCbCr data to further improve picture quality.

Macroblocks

With H.264, the partitioning of the 16 × 16 macroblocks has been extended. Such fine granularity leads to a potentially large number of motion vectors per macroblock (up to 32) and number of blocks that must be interpolated (up to 96). To constrain encoder/decoder complexity, there are limits on the number of motion vectors used for two consecutive macroblocks.

Error concealment is improved with *Flexible Macroblock Ordering* (FMO), which assigns macroblocks to another slice so they are transmitted in a non-scanning sequence. This reduces the chance that an error will affect a large spatial region, and improves error concealment by being able to use neighboring macroblocks for prediction of a missing macroblock.

Motion Compensation

¼-Pixel Motion Compensation:

Motion compensation accuracy is improved from the ½-pixel accuracy used by most earlier video codecs. H.264 supports the same ¼-pixel accuracy that is used on the latest MPEG-4 video codec.

Multiple Reference Frames

H.264 adds supports for multiple reference frames. This increases compression by improving the prediction process and increases error resilience by being able to use another reference frame in the event that one was lost.

A single macroblock can use up to 8 reference frames (up to 3 for HDTV), with a total limit of 16 reference frames used within a frame.

To compensate for the different temporal distances between current and reference frames, predicted blocks are averaged with configurable weighting parameters. These parameters can either be embedded within the bitstream or the decoder may implicitly derive them from temporal references.

Transform, Scaling, and Quantization

H.264 uses a simple 4 × 4 integer transform. In contrast, older video codecs use an 8 × 8 DCT that operates on floating-point coefficients. An additional

2×2 transform is applied to the four CbCr DC coefficients. Intra 16×16 macroblocks have an additional 4×4 transform performed for the sixteen Y DC coefficents.

Blocking and ringing artifacts are reduced as a result of the smaller block size used by H.264. The use of integer coefficients eliminates rounding errors that cause drifting artifacts common with DCT-based video codecs.

For quantization, H.264 uses a set of 52 uniform scalar quantizers, with a step increment of about 12.5% between each.

The quantized coefficients are then scanned, from low frequency to high frequency, using one of two scan orders.

Entropy Coding

After quantization and zig-zag scanning, H.264 uses two types of entropy encoding: variable-length coding (VLC) and Context Adaptive Binary Arithmetic Coding (CABAC).

For everything but the transform coefficients, H.264 uses a single Universal VLC (UVLC) table that uses an infinite-extend codeword set (Exponential Golomb). Instead of multiple VLC tables as used by other video codecs, only the mapping to the single UVLC table is customized according to statistics.

For transform coefficients, which consume most of the bandwidth, H.264 uses Context-Adaptive Variable Length Coding (CAVLC). Based upon previously processed data, the best VLC table is selected.

INSTANT SUMMARY

The MPEG compression standards have contributed greatly to the proliferation of video in today's devices. This chapter covered the major features of the most common standards:

- Compression definitions
- MPEG-1
- MPEG-2
- MPEG-4
- H.264

Chapter 8

Digital Television (DTV)

In an Instant

- Definitions
- ATSC digital television
- OpenCable™ digital television
- DVB digital television
- ISDB digital television

Definitions

The *ATSC* (Advanced Television Systems Committee) digital television (DTV) broadcast standard is used in the United States, Canada, South Korea, Mexico, and Argentina.

The three other primary DTV standards are *DVB* (Digital Video Broadcast), *ISDB* (Integrated Services Digital Broadcasting), and *OpenCable*™. The basic audio and video capabilities are very similar. The major differences are the RF modulation schemes and the level of definition for non-audio/video services.

QAM (*quadrature amplitude modulation*) is a method of encoding digital data onto a carrier for RF transmission. It is typically used for cable transmission of DTV signals.

ATSC DIGITAL TELEVISION

A comparison of the ATSC standards is shown in Table 8.1.

The ATSC standard is actually a group of standards:

A/52 Digital Audio Compression (AC-3 and E-AC-3) Standard
A/53 ATSC Digital Television Standard
A/57 Content Identification and Labeling for ATSC Transport
A/64 Transmission Measurement and Compliance for Digital Television

TABLE 8.1 Comparison of ATSC Standards

Parameter	ATSC-T (Terrestrial)	ATSC-C (Cable)	ATSC-S (Satellite)	ATSC-T E-VSB (Terrestrial)
Video compression	MPEG-2			MPEG-2, MPEG-4.10 (H.264)
Audio compression	Dolby® Digital			Dolby® Digital, Dolby® Digital Plus
Multiplexing	MPEG-2 transport stream			
Modulation	8-VSB	16-VSB[1]	QPSK, 8PSK	uses ATSC-T
Channel bandwidth	6 MHz	6 MHz	–	–

Note: 1. Most digital cable systems use QAM instead of 16-VSB.

A/65	Program and System Information Protocol for Terrestrial Broadcast and CableA/70 Conditional Access System for Terrestrial Broadcast
A/76	Programming Metadata Communication Protocol
A/80	Modulation and Coding Requirements for Digital TV (DTV) Applications Over Satellite
A/81	Direct-to-Home Satellite Broadcast Standard
A/90	Data Broadcast Standard
A/92	Delivery of IP Multicast Sessions over Data Broadcast Standard
A/93	Synchronized/Asynchronous Trigger Standard
A/94	Data Application Reference Model
A/95	Transport Stream File System Standard
A/96	Interaction Channel Protocols
A/97	Software Download Data Service
A/100	DTV Application Software Environment: Level 1 (DASE-1)
A/101	Advanced Common Application Platform (ACAP)
A/110	Synchronization Standard for Distributed Transmission

The ATSC standard uses an MPEG-2 transport stream to convey compressed digital video, compressed digital audio, and data over a single 6-MHz channel. Multiple video streams, multiple audio streams, and/or data may be present in the MPEG-2 transport stream. For example, both HD and SD versions of a program may be present, along with data, such as a local weather forecast.

The MPEG-2 transport stream has a maximum bit-rate of ~19.4 Mbps (6-MHz over-the-air channel) or ~38.8 Mbps (6-MHz digital cable channel).

FAQs

Can the bit-rate be divided?

The 19.4 Mbps bit-rate can be used in a very flexible manner, trading off the number of programs offered versus video quality and resolution. For example,

(1) HDTV program
(1) HDTV program + (1) SDTV program + data
(4) SDTV programs

Video Capability

Although any resolution may be used as long as the maximum bit-rate is not exceeded, there are several standardized resolutions. Both interlaced and progressive pictures are permitted for most of the resolutions.

Video compression is based on MPEG-2. However, there are some minor constraints on some of the MPEG-2 parameters, as discussed in Chapter 7. Support for using MPEG-4.10 (H.264) up to HP@L4.0 is being added to the specifications.

Audio Capability

Audio compression is implemented using Dolby® Digital and supports 1–5.1 channels.

The main audio, or associated audio which is a complete service (containing all necessary program elements), has a bit-rate ≤448 kbps (384 kbps is typically used). A single channel associated service containing a single program element has a bit-rate ≤128 kbps. A two channel associated service containing only dialogue has a bit-rate ≤192 kbps. The combined bit-rate of a main and associated service which are intended to be decoded simultaneously must be ≤576 kbps.

Program and System Information Protocol (PSIP)

Enough bandwidth is available within the MPEG-2 transport stream to support several low-bandwidth non-television services such as program guide, closed captioning, weather reports, stock indices, headline news, software downloads, pay-per-view information, etc. The number of additional non-television services (virtual channels) may easily reach ten or more. In addition, the number and type of service will constantly be changing.

To support these non-television services in a flexible yet consistent manner, the Program and System Information Protocol (PSIP) was developed. PSIP is a small collection of hierarchically associated tables (see Figure 8.1 and Table 8.2) designed to extend the MPEG-2 PSI tables. It describes the information for all virtual channels carried in a particular MPEG-2 transport stream. Additionally, information for analog broadcast channels may be incorporated.

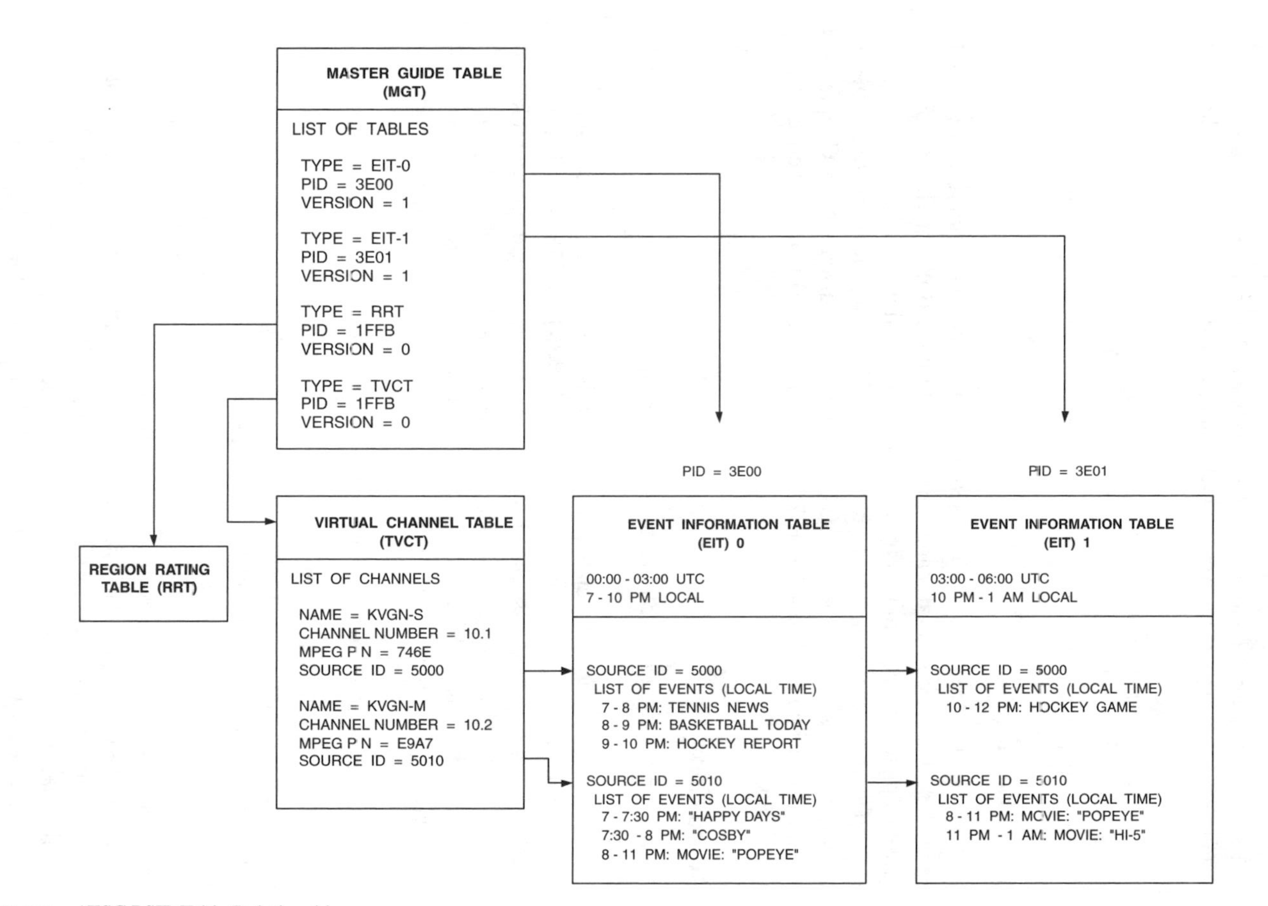

FIGURE 8.1 ATSC PSIP Table Relationships.

TABLE 8.2 List of ATSC PSIP Tables, Descriptors, and Descriptor Locations

Descriptor	Descriptor Tag	Terrestrial Broadcast Tables									
		PMT	MGT	VCT	RRT	EIT	ETT	STT	DCCT	DCCSCT	CAT
PID		per PAT	0x1FFB	0x1FFB	0x1FFB	per MGT	per MGT	0x1FFB	0x1FFB	0x1FFB	0x0001
Table_ID		0x02	0xC7	0xC8	0xCA	0xCB	0xCC	0xCD	0xD3	0xD4	0x80, 0x81 (ECM) 0x82 – 0x8F (EMM)
Repetition rate		400 ms	150 ms	400 ms	1 min	0.5 sec	1 min	1 sec	400 ms	1 hour	
AC-3 audio stream	1000 0001	M				M					
ATSC CA	1000 1000			O		O					
ATSC private information*	1010 1101										
CA	0000 1001	M									M
Caption service	1000 0110	M				M					
Component name	1010 0011	M									
Content advisory	1000 0111	M				M					
Content identifier	1011 0110	O				M					
DCC arriving request	1010 1001								M		
DCC departing request	1010 1000								M		

(Continued)

TABLE 8.2 (Continued)

Descriptor	Descriptor Tag	Terrestrial Broadcast Tables									
		PMT	MGT	VCT	RRT	EIT	ETT	STT	DCCT	DCCSCT	CAT
Enhanced signaling	1011 0010	M PMT-E									
Extended channel name	1010 0000			M							
Genre	1010 1011					M					
Redistribution control	1010 1010	M				M					
Service location	1010 0001			M							
SRM reference	0000 1001										M
Stuffing*	1000 0000										
Time-shifted service	1010 0010			M							

*Note: 1. M = when present, required in this table. O = may be present in this table also. * = no restrictions.*

Required Tables

Event Information Table (EIT)

There are up to 128 EITs, EIT-0 through EIT-127, each of which describes the events or TV programs associated with each virtual channel listed in the VCT. Each EIT is valid for three hours. Since there are up to 128 EITs, up to 16 days of programming may be advertised in advance. The first four EITs are required (the first 24 are recommended) to be present.

Information provided by the EIT includes start time, duration, title, pointer to optional descriptive text for the event, advisory data, caption service data, audio service descriptor, and so on.

Master Guide Table (MGT)

This table provides general information about the other tables. It defines table sizes, version numbers, and packet identifiers (PIDs).

Rating Region Table (RRT)

This table transmits the rating system, commonly referred to as the "V-chip."

System Time Table (STT)

This table serves as a reference for the time of day. Receivers use it to maintain the correct local time.

Terrestrial Virtual Channel Table (TVCT)

This table, also referred to as the VCT although there is also a Cable VCT (CVCT) and Satellite VCT (SVCT), contains a list of all the channels in the transport stream that are or will be available, plus their attributes. It may also include the broadcaster's analog channel and digital channels in other transport streams.

Attributes for each channel include major/minor channel number, short name, Transport/Transmission System ID (TSID) that uniquely identifies each station, etc. The Service Location Descriptor is used to list the PIDs for the video, audio, data, and other related elementary streams.

Optional Tables

Extended Text Table (ETT)

For text messages, there can be several ETTs, each having its PID defined by the MGT. Messages can describe channel information, coming attractions, movie descriptions, and so on.

Directed Channel Change Table (DCCT)

The DCCT contains information needed for a channel change to be done at a broadcaster-specified time. The requested channel change may be unconditional or may be based upon criteria specified by the viewer.

Directed Channel Change Selection Code Table (DCCSCT)

The DCCSCT permits a broadcast program categorical classification table to be downloaded for use by some Directed Channel Change requests.

Descriptors

Much like MPEG-2, ATSC uses descriptors to add new functionality. In addition to various MPEG-2 descriptors, one or more of these ATSC-specific descriptors may be included within the PMT or one or more PSIP tables to extend data within the tables. A descriptor not recognized by a decoder must be ignored by that decoder. This enables new descriptors to be implemented without affecting receivers that cannot recognize and process the descriptors.

AC-3 Audio Stream Descriptor

This ATSC descriptor indicates Dolby® Digital or Dolby® Digital Plus audio is present.

ATSC CA Descriptor

This ATSC descriptor has a syntax almost the same as the MPEG-2 CA descriptor.

ATSC Private Information Descriptor

This ATSC descriptor provides a way to carry private information. More than one descriptor may appear within a single descriptor.

Component Name Descriptor

This ATSC descriptor defines a variable-length text-based name for any component of the service.

Content Advisory Descriptor

This ATSC descriptor defines the ratings for a given program.

Content Identifier Descriptor

This ATSC descriptor is used to uniquely identify content with the ATSC transport.

DCC Arriving Request Descriptor

This ATSC descriptor provides instructions for the actions to be performed by a receiver upon arrival to a newly changed channel:

Display text for at least 10 seconds, or for a less amount of time if the viewer issues a "continue," "OK," or equivalent command.

Display text indefinitely, or until the viewer issues a "continue," "OK," or equivalent command.

DCC Departing Request Descriptor

This ATSC descriptor provides instructions for the actions to be performed by a receiver prior to leaving a channel:

Cancel any outstanding things and immediately perform the channel change.

Display text for at least 10 seconds, or for a smaller amount of time if the viewer issues a "continue," "OK," or equivalent command.

Display text indefinitely, or until the viewer issues a "continue," "OK," or equivalent command.

Enhanced Signaling Descriptor

This ATSC descriptor identifies the terrestrial broadcast transmission method of a program element.

Extended Channel Name Descriptor

This ATSC descriptor provides a variable-length channel name for the virtual channel.

Genre Descriptor

This ATSC descriptor provides genre, program type, or category information for events., and may appear in the descriptor() loop for the given EIT event. It references entries in the Categorical Genre Code Assignments Table and may include references to expansions to that table provided by the DCC Selection Code.

Redistribution Control Descriptor

This ATSC descriptor conveys any redistribution control information held by the program rights holder for the content.

Service Location Descriptor

This ATSC descriptor specifies the stream type, PID, and language code for each elementary stream. It is present in the VCT for each active channel.

SRM Reference Descriptor

This ATSC descriptor is a specific implementation of the MPEG-2 CA Descriptor. It is used to signal that a System Renewability Message is present for the System Renewability Message Table (SRMT). It is present in the CAT.

Time-Shifted Service Descriptor

This ATSC descriptor links one virtual channel with up to 20 other virtual channels carrying the same programming, but time-shifted. A typical application is for Near Video On Demand (NVOD) services.

Application Block Diagrams

Figure 8.2 illustrates a typical ATSC receiver set-top box block diagram. A common requirement is the ability to output both high-definition and standard-definition versions of a program simultaneously.

Figure 8.3 illustrates a typical ATSC digital television block diagram. A common requirement is the ability to decode two programs simultaneously to support Picture-in-Picture (PIP).

OPENCABLE™ DIGITAL TELEVISION

OpenCable™ is a digital cable standard for the United States, designed to offer interoperability between different hardware and software suppliers. A subset of the standard is being incorporated inside digital televisions.

A summary of the OpenCable™ standard is shown in Table 8.3.

OpenCable™ receivers use the following four communications channels over the digital cable network:

6-MHz NTSC analog channels. They are typically located in the 54–450 MHz range. Each channel carries one program.

6-MHz Forward Application Transport (FAT) channels, which carry content via MPEG-2 transport streams. They use QAM encoding and are typically located in the 450–864 MHz range. Each channel can carry multiple programs.

Out-of-Band (OOB) Forward Data Channels (FDC). They use QPSK modulation and are typically located in the 70–130 MHz range, spaced between the 6 MHz NTSC analog and/ or FAT channels. SCTE 55-1 and SCTE 55-2 are two alternative implementations.

Out-of-Band (OOB) Reverse Data Channels (RDC). They use QPSK modulation and are typically located in the 5–42 MHz range. SCTE 55-1, SCTE 55-2, and DOCSIS® provide three alternative implementations.

OpenCable™ receivers obtain content by tuning to one of many 6-MHz channels available via the cable TV connection. When the selected channel is a legacy analog channel, the signal is processed using a NTSC audio/video/VBI decoder. When the selected channel is a digital channel, it is processed by a QAM demodulator and then a CableCARD™ for content descrambling (conditional access descrambling). The conditional access descrambling is specific to a given cable system and is usually proprietary. The CableCARD™ then rescrambles the content to a common algorithm and passes it on to the MPEG-2 decoder.

Insider Info

The multi-stream CableCARD™ is capable of handling up to six different channels simultaneously, enabling picture-in-picture and DVR (digital video recording) capabilities.

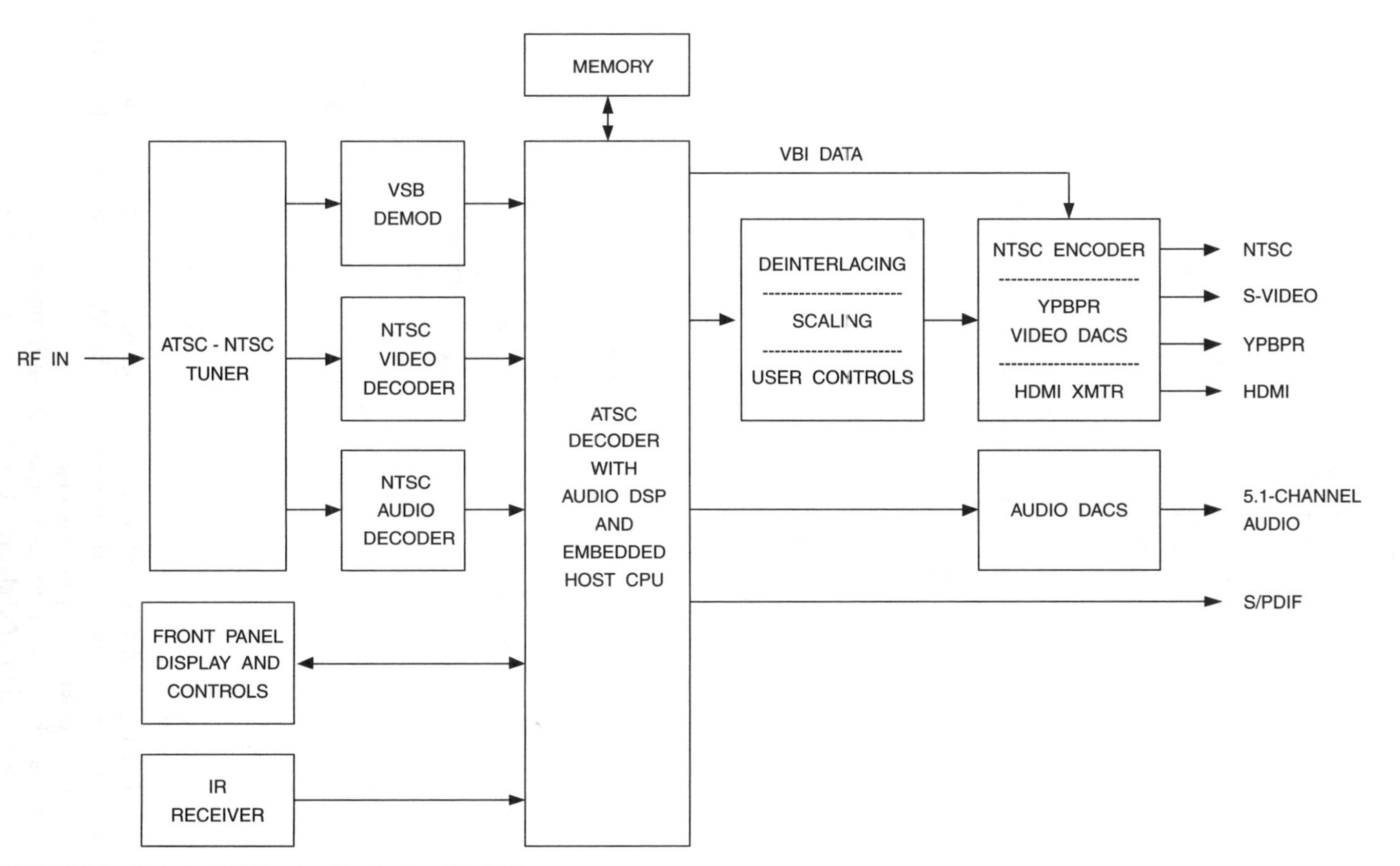

FIGURE 8.2 Typical ATSC Receiver Set-Top Box Block Diagram.

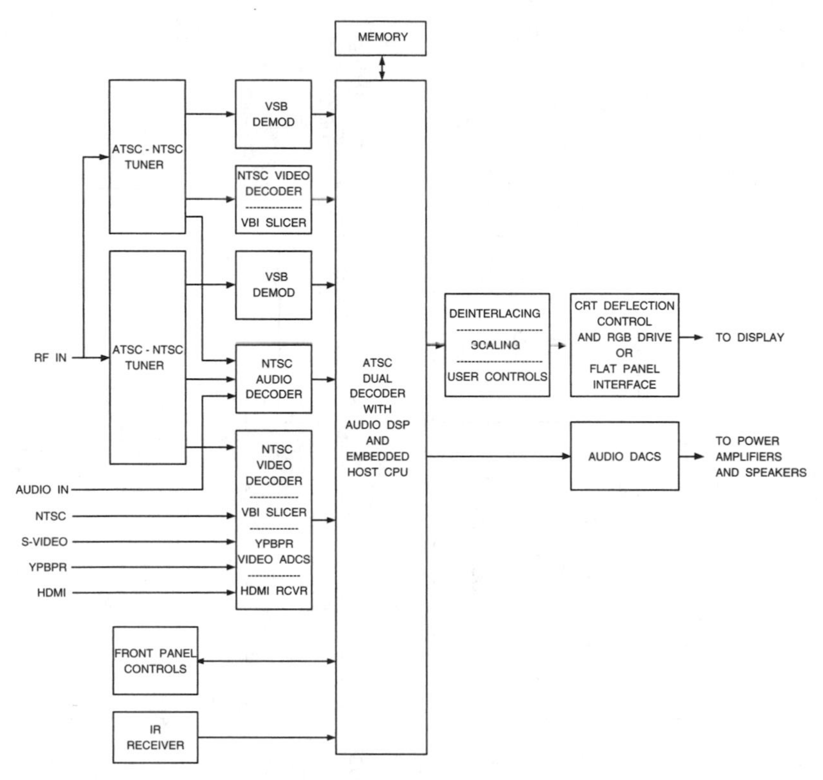

FIGURE 8.3 Typical ATSC Digital Television Block Diagram.

TABLE 8.3 Summary of the OpenCable™ Standard

Parameter	OpenCable™
Video compression	MPEG-2
Audio compression	Dolby® Digital
Multiplexing	MPEG-2 transport stream
Modulation	QAM
Channel bandwidth	6 MHz

When the CableCARD™ is not inserted, the output of the digital tuner's QAM demodulator is routed directly to the MPEG-2 decoder. However, encrypted content will not be viewable.

OpenCable™ receivers also obtain control information and other data by tuning to the OOB FDC channel. Using a dedicated tuner, the receiver remains tuned to the OOB FDC to receive information continuously. This information is also passed to the CableCARD™ and MPEG-2 decoder for processing.

The bidirectional OpenCable™ receiver can also transmit data using the OOB RDC.

The OpenCable™ standard uses an MPEG-2 transport stream to convey compressed digital video, compressed digital audio, and ancillary data over a single 6-MHz FAT channel. Multiple video streams, multiple audio streams and/or data may be present in the MPEG-2 transport stream.

The MPEG-2 transport stream has a constant bit-rate of ~27 Mbps (64-QAM modulation), ~38.8 Mbps (256-QAM), or ~44.3 Mbps (1024-QAM).

The available bit-rate can be used in a very flexible manner, trading off the number of programs offered versus video quality and resolution. For example, if MPEG-2 video, statistical multiplexing, and 256-QAM are used:

(4) HDTV programs
(2) HDTV programs + (6) SDTV programs + data
(18) SDTV programs

Video Capability

Digital video compression is implemented using MPEG-2 and has the same requirements as ATSC. There are some minor constraints on some of the MPEG-2 parameters, as discussed in Chapter 7. Support for using MPEG-4.10 (H.264) up to HP@L4.0 is being added to the specifications.

Although any resolution may be used as long as the maximum bit-rate is not exceeded, there are several standardized resolutions. Both interlaced and progressive pictures are permitted for most of the resolutions.

Compliant receivers must also be capable of tuning to and decoding analog NTSC signals.

Audio Capability

Digital audio compression is implemented using Dolby® Digital and has the same requirements as ATSC. Compliant receivers must also be capable of decoding the audio portion of analog NTSC signals.

Application Block Diagrams

Figure 8.4 illustrates an OpenCable™ set-top box. Part of the requirements is the ability to output both high-definition and standard-definition versions of HD content simultaneously.

DVB DIGITAL TELEVISION

The DVB (Digital Video Broadcast) digital television (DTV) broadcast standard is used in most regions except the United States, Canada, South Korea, Taiwan, Brazil, and Argentina.

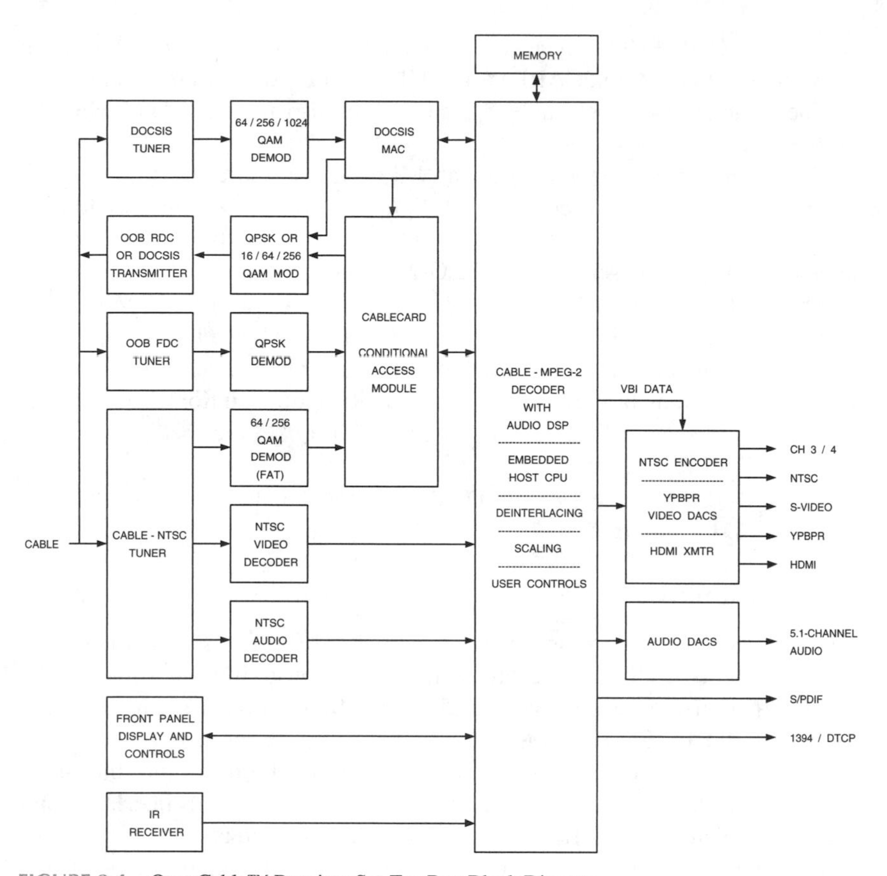

FIGURE 8.4 OpenCable™ Receiver Set-Top Box Block Diagram.

The DVB standard is actually a group of ETSI standards. A comparison of the DVB standards is given in Table 8.4. DVB uses a MPEG-2 transport stream to convey compressed digital video, compressed digital audio, and data over a 6-, 7- or 8-MHz channel. Multiple video streams, multiple audio streams, and/ or data may be present in the MPEG-2 transport stream.

The MPEG-2 transport stream has a maximum bit-rate of ~24.1 Mbps (8 MHz DVB-T) or ~51 Mbps (8-MHz 256-QAM DVB-C). DVB-S bit-rates are dependent on the transponder bandwidth and code rates used, and can approach 54 Mbps (DVB-S2 offers a 25–35% bit-rate capacity gain over DVB-S). The bit-rate can be used in a very flexible manner, trading off the number of programs offered versus video quality and resolution.

DVB-H and DVB-SH for mobile applications use IP datacasting within DVB-T and DVB-S, respectively. RTP-encapsulated MPEG-2 transport streams are used.

TABLE 8.4 Comparison of DVB Standards

Parameter	DVB-T (Terrestrial)	DVB-C (Cable)	DVB-S/-S2 (Satellite)	DVB-H (Handheld)	DVB-SH (Handheld)
Video compression	MPEG-2, MPEG-4.10 (H.264), SMPTE 421M (VC-1)			MPEG-4.10 (H.264), SMPTE 421M (VC-1)	
Audio compression	MPEG, Dolby® Digital, Dolby® Digital Plus, DTS®, MPEG-4 AAC, MPEG-4 HE-AAC v1/v2			MPEG-4 AAC, MPEG-4 HE-AAC v1/v2, AMR-WB+	
Multiplexing	MPEG-2 transport stream			RTP-encapsulated MPEG-2 transport stream	
Modulation	COFDM	QAM	QPSK	uses DVB-T	uses DVB-S
Channel bandwidth	6, 7, or 8 MHz	6, 7, or 8 MHz	–	6, 7, or 8 MHz	–

Second generation versions of DVB-T and DVB-C (called DVB-T2 and DVB-C2, respectively) are being investigated.

Insider Info

Much like MPEG-2, DVB uses descriptors to add new functionality. In addition to various MPEG-2 descriptors, one or more of numerous DVB-specific descriptors may be included to extend data within the system information tables.

Video Capability

Although any resolution may be used as long as the maximum bit-rate is not exceeded, there are several standardized resolutions. Both interlaced and progressive pictures are permitted for most of the resolutions.

DVB-T, DVB-C, DVB-S, and DVB-S2 support MPEG-2 (MP@ML, MP@HL), MPEG-4.10 (MP@L3, HP@L4), and SMPTE 421M (AP@L1, AP@L3) video.

DVB-IP ("DVB over IP", used by DVB-H, DVB-SH and DVB-IPTV) adds additional support for MPEG-4.10 (BP@L1b, BP@L1.2, BP@L2) and SMPTE 421M (SP@LL, SP@ML, AP@L0) video.

Audio Capability

DVB-T, DVB-C, DVB-S, and DVB-S2 support MPEG-1 Layer II, MPEG-2 BC multi-channel Layer II, Dolby® Digital, Dolby® Digital Plus, DTS®, MPEG-4 AAC, and MPEG-4 HE-AAC v1/v2 audio.

DVB-IP ("DVB over IP". used by DVB-H, DVB-SH and DVB-IPTV) adds additional support for AMR-WB+ audio.

Application Block Diagrams

Figures 8.5 and 8.6 illustrate a typical DVB-S set-top box.

ISDB DIGITAL TELEVISION

The ISDB (Integrated Services Digital Broadcasting) digital television (DTV) broadcast standard is used in Japan.

ISDB builds on DVB, adding additional services required for Japan. A comparison of the ISDB standards is shown in Table 8.5.

The ISDB standard is actually a group of ARIB standards. ISDB uses an MPEG-2 transport stream to convey compressed digital video, compressed digital audio, and data. Like DVB, this transport stream is then transmitted either via terrestrial, cable, or satellite. Interactive applications are based on BML (Broadcast Mark-up Language).

ISDB-S (Satellite)

Two satellite standards exist, ISDB-S, also known as the BS (broadcast satellite) system, and DVB-S, also known as the CS (communications satellite) system.

ISDB-S (BS) is also specified by ITU-R BO.1408. It has a maximum bit-rate of ~52.2 Mbps using TC8PSK modulation and a 34.5 MHz transponder.

CS supports only one transport stream per transport channel, and supports up to ~34 Mbps on a 27 MHz channel. The modulation scheme is QPSK, just like DVB-S. Unlike the other variations of ISDB, CS uses MPEG-2 MP@ML video (480i or 480p) and MPEG-2 BC audio.

ISDB-C (Cable)

ISDB-C uses 64-QAM modulation, with two versions: one that supports only a single transport stream per transmission channel and one that supports multiple transport streams per transmission channel. On a 6-MHz channel, ISDB-C can transmit up to ~29.16 Mbps. As the bit-rate on an ISDB-S satellite channel is 2× that, two cable channels can be used to rebroadcast satellite information.

ISDB-C also allows passing through OFDM-based ISDB-T signals since the channel bandwidths are the same.

ISDB-S (BS) signals can also be passed through by downconverting the satellite signals at the cable head-end, then up-converting them at the receiver. This technique is suitable only for cable systems that have many (up to 29) unused channels.

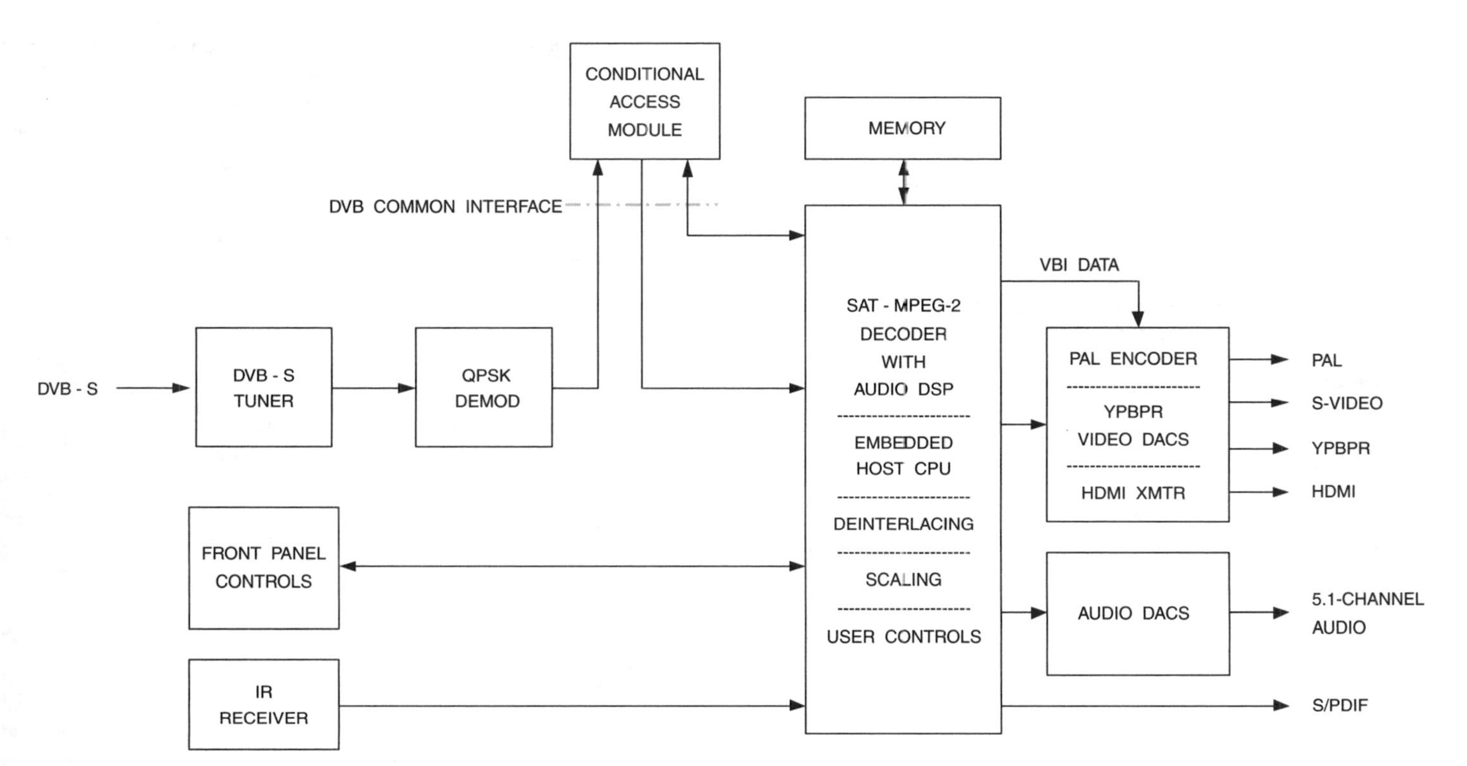

FIGURE 8.5 DVB Receiver Set-Top Box Block Diagram (Multicrypt).

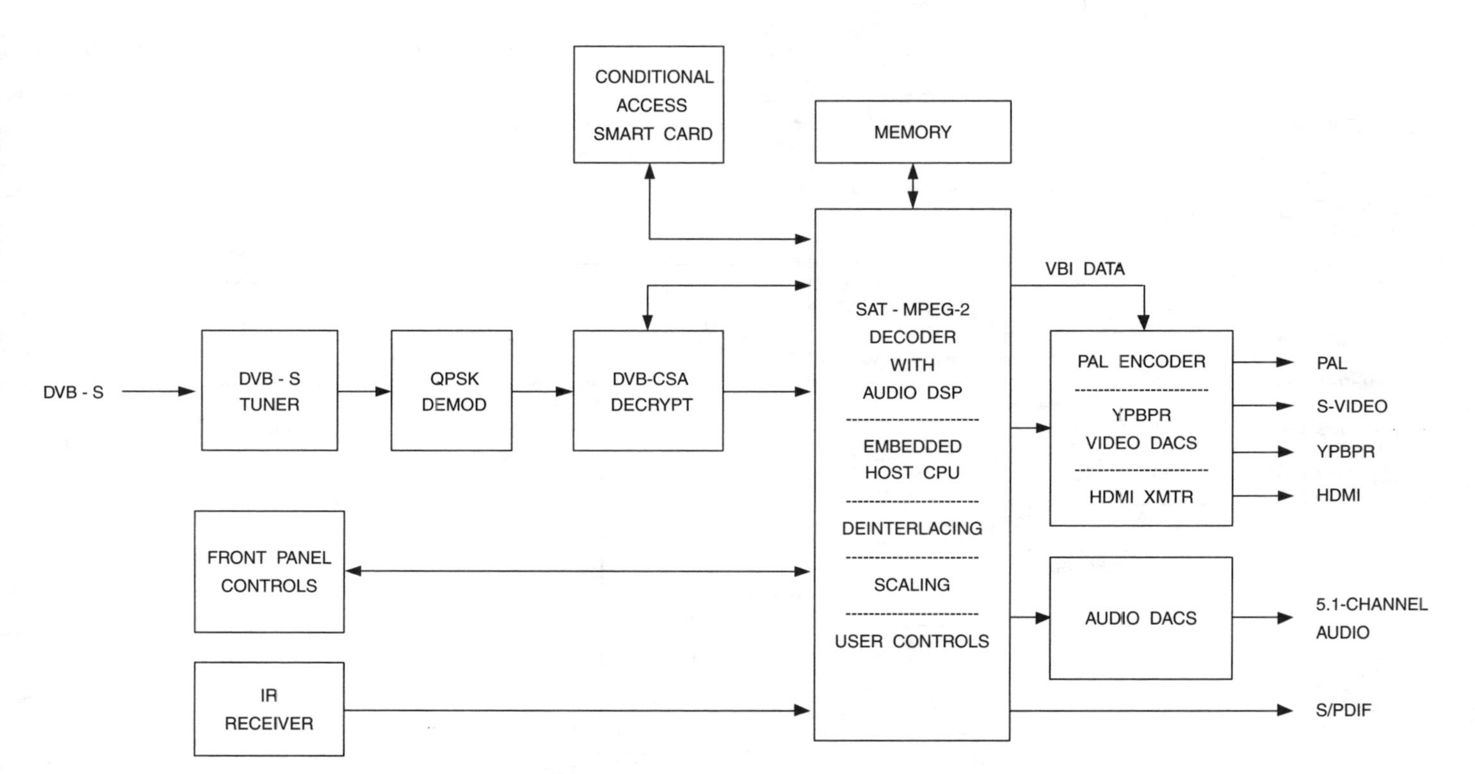

FIGURE 8.6 DVB Receiver Set-Top Box Block Diagram (Simulcrypt).

TABLE 8.5 Comparison of ISDB Standards

Parameter	ISDB-T (Terrestrial)	ISDB-C (Cable)	ISDB-S (Satellite)
Video compression	MPEG-2, MPEG-4.10 (H.264)		
Audio compression	MPEG-2 AAC, MPEG-4 HE-AAC		
Multiplexing	MPEG-2 transport stream		
Modulation	BST-OFDM[1]	QAM	PSK
Channel bandwidth	6, 7, or 8 MHz	6, 7, or 8 MHz	–

Note: 1. BST-OFDM = Bandwidth segmental transmission of OFDM.

ISDB-T (Terrestrial)

ISDB-T, the terrestrial broadcast standard, is also specified by ITU-R BT.1306. It has a maximum bit-rate of ~23.2 Mbps using 5.6 MHz of bandwidth. ISDB-T also supports bandwidths of 6, 7, and 8 MHz.

The bandwidth is divided into 13 OFDM segments; each segment can be divided in up to three segment groups (hierarchical layers) having different transmission parameters such as the carrier modulation scheme, inner-code coding rate, and time interleaving length. This enables the same program to be broadcast in different resolutions, allowing a mobile receiver to show a standard-definition picture while a stationary receiver shows a high-definition picture.

Video Capability

There are several standardized resolutions, indicated in Table 18.6.

Primary video compression is based on MPEG-2 MP@ML or MP@HL. However, there are some minor constraints on some of the MPEG-2 parameters, as discussed within Chapter 7.

MPEG-4.2 Simple Profile or Core Profile video is also supported, using resolutions of 176 × 144 (64 or 384 kbps) or 325 × 288 (128, 384, or 2000 kbps).

MPEG-4.10 (H.264) Baseline Profile or Main Profile video is also supported, using resolutions of 176 × 144 (64 kbps) or 325 × 288 (192, 384, 768, 2000, or 4000 kbps).

TABLE 8.6 Common Active Resolutions for ISDB Digital Television

Active Resolution (Y)	SDTV or HDTV	Frame Rate (p = progressive, i = interlaced)				MPEG-2	MPEG-4.2	MPEG-4.10 (H.264)
		23.976p 24p	29.97i 30i	29.97p 30p	59.94p 60p			
176 × 120	SDTV	×		×	×	×		
176 × 144		×		×	×	×	×	×
352 × 240		×		×	×	×		
352 × 288		×		×	×	×	×	×
352 × 480			×			×		
480 × 480			×			×		
544 × 480			×			×		
720 × 480			×			×		
1280 × 720	HDTV	×		×	×	×		
1440 × 1080		×	×	×		×		
1920 × 1080		×	×	×		×		

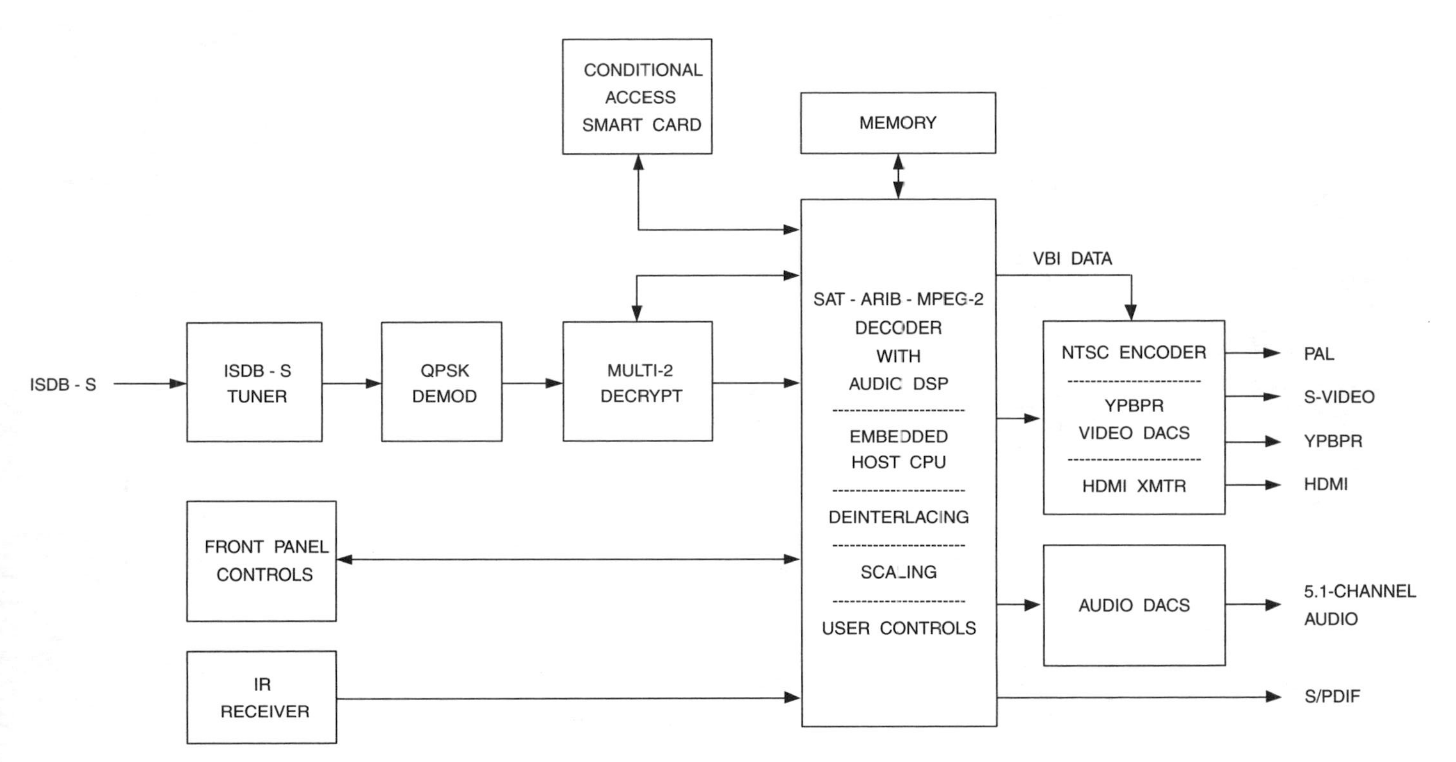

FIGURE 8.7 ISDB Receiver Set-Top Box Block Diagram.

Insider Info

Much like MPEG-2, ISDB uses descriptors to add new functionality. One or more ISDB specific descriptors can be included to extend data within the table. A descriptor not recognized by a decoder must be ignored by that decoder; this allows new descriptors to be implemented without affecting receivers that cannot recognize and process the description.

Audio Capability

Primary audio compression is implemented using MPEG-2 AAC-LC with up to 5.1 channels. ISDB also supports MPEG-4 HE-AAC audio.

Application Block Diagrams

Figure 8.7 illustrates a typical ISDB-S set-top box.

INSTANT SUMMARY

Digital television (DTV) is made up of several different international broadcast standards. This chapter covers the following systems, providing information on each and a comparison of their features:

- ATSC
- OpenCable
- DVB
- ISDB

Index

0 IRE blanking pedestal, 67, 69
¼-pixel motion compensation, 198
2D Audio Graphics profile, 192
3D Audio Graphics profile, 192
4:2:0 YCbCr color space, 189
6-MHz Forward Application Transport (FAT) channels, 210
6-MHz NTSC analog channels, 210
7.5 IRE blanking pedestal, 67, 68, 69
16 × 8 motion compensation option, 186
25-pin parallel interface, 86, 87, 88
27 MHz parallel interface, 88–9, 90
36 MHz parallel interface, 89, 90, 91
74.25 and 74.176 MHz parallel interface, 91–2, 93
93-pin parallel interface, 91, 92
148.5 and 148.35 MHz parallel interface, 93–4
480i and 480p system:
  interlaced analog component video, 35–6
  interlaced analog composite video, 35
  interlaced digital component video, 36–40
  progressive analog component video, 36
  progressive digital component video, 40–1, 42, 43
480i system, definition of, 7
480p system, definition of, 7
576i and 576p system:
  interlaced analog component video, 44
  interlaced analog composite video, 41–3
  interlaced digital component video, 45–7
  progressive analog component video, 44
  progressive digital component video, 47–8, 49–51
576i system, definition of, 7
576p system, definition of, 7
720p system:
  progressive analog component video, 48, 51
  progressive digital component video, 51–5
1080i and 1080p system:
  interlaced analog component video, 54
  interlaced digital component video, 55–6, 57, 58
  progressive analog component video, 54–5
  progressive digital component video, 56, 57, 58–60

A

AC-3 audio stream descriptor, 208
Active video, 31
Adaptive contrast enhancement, *see* Dynamic contrast
Advanced Simple objects, 190
Alpha mixing, 125, 130–1
Analog video interface:
  D-connector interface, 75, 77, 78, 79
  HDTV RGB interface, 67–8, 70
  HDTV YPbPr interface, 71–2, 74, 75, 76, 77
  pro-video analog interfaces, 77–8, 80, 81, 82, 83
  S-video interface, 64–5
  SCART interface, 65, 66, 67

Analog video interface (*continued*)
SDTV RGB interface, 65, 67, 68, 69
SDTV YPbPr interface, 69–71, 72, 73, 74
VGA interface, 78, 83, 84
Ancillary data:
pro-video component interfaces, 86, 88
pro-video composite interfaces, 100–1, 102–3
Anti-aliased resampling, 136–7
Aspect ratio, 31
ATSC digital television, 201
audio capability, 203
block diagram, 210, 211, 212
CA descriptor, 208
private information descriptor, 208
program and system information protocol, 203–9
video capability, 203
Audio and video compression, 7–8
Audio capability, 203
ATSC digital television, 203
DVB digital television, 215–16
ISDB digital television, 222
OpenCable™, 213
Audio coding:
MPEG-1, 168–9
MPEG-2, 175

B

B (bidirectional) frames, 171–2
B (bidirectional) pictures, 185, 187
Baseline profile (BP), 197
BIFS (BInary Format for Scenes), 192
2D primitives, 195
audio enhancement, 195
compressed binary format, 194
facial animation, 195
streamed animation, 195
streaming, 194
Blanking level, 32
Blue stretch, 129
Brightness control, 126, 127
BT.601 video interface, 35, 36, 37, 45, 108
BT.656 interface, 88, 107, 109
BT.1358, 40–1, 42, 43, 47

C

Cable VCT, 207
CableCARD™, 210, 213
Chroma keying, 125, 131–2, 133, 134
Chromaticity diagram, 22–5
CIE 1931 chromaticity diagram, 23, 24
Coded frames, 171–2
Coded pictures, 185–6
Coding ranges, 33
Color bars, 32
Color burst, 151, 154
Color correction, 130
Color information:
NTSC, 152–3
PAL, 157
SECAM, 161
Color modulation:
NTSC, 153–4
PAL, 157–8
SECAM, 161
Color space, 1, 4
chromaticity diagram, 22–5
gamma correction, 26–9
HLS color space, 19
HSI color space, 19–22
HSV color space, 19–22
non-RGB color space considerations, 25–6
RGB color space, 15–16
YCbCr color space, 18–19
YIQ color space, 17–18
YUV color space, 16–17
Color transient improvement, 126, 128
Component name descriptor, 208
Component video, 1
Composite video, definition of, 1
Composite video generation:
NTSC, 154–5
PAL system, 158–9
SECAM, 161

Constrained Parameters Bitstream (CPB), 169, 170, 176
Consumer component interfaces:
- Digital Visual Interface (DVI), 110–14
- Gigabit Video Interface (GVIF), 117–19
- High-Definition Multimedia Interface (HDMI), 114–17
- Open LVDS Display Interface (OpenLDI), 117, 118

Consumer transport interfaces:
- digital camera specification, 122–3
- Digital Transmission Content Protection (DTCP), 122
- Ethernet, 120
- IEEE 1394, 120–1, 122
- USB 2.0, 119–20

Content advisor descriptor, 208
Content identifier descriptor, 208
Context Adaptive Binary Arithmetic Coding (CABAC), 199
Context Adaptive Variable Length Coding (CAVLC), 199
Contrast control, 126
Core objects, 190
CRT display, 26, 27

D

D (DC) frames, 172
D (DC) pictures, 185
D-connector interface, 75, 77, 78, 79
Data partitioning, 180
Db information, 161
DCT (Discrete Cosine Transform)-based compression, 147–50
- quantization, 149
- run length coding, 150
- variable-length coding, 150
- zig-zag scanning, 149

Descrambler circuit, 95
Digital-analog (DVI-I) connector, 113–14, 115
Digital camera specification, 122–3
Digital component video, 32
- coding ranges, 33
- EDTV sample rate selection, 35
- HDTV sample rate selection, 35
- SDTV sample rate selection, 33–5

Digital Display Working Group (DDWG), 110
Digital Flat Panel (DFP) interface, 110, 116
Digital media adapters, 2, 9
Digital-only (DVI-D) connector, 106, 112, 114
Digital set-top boxes, 2
Digital Storage Media Command and Control (DSM-CC), 175
Digital television (DTV), 2
- ATSC, 201–10
- DVB (Digital Video Broadcast), 213–16, 217, 218
- ISDB, 216, 219–22
- OpenCable™, 210, 212, 213, 214
- set-top boxes, 10, 11

Digital Transmission Content Protection (DTCP), 122
Digital Transmission Licensing Administrator (DTLA), 122
Digital Video Broadcasting (DVB), 201
- audio capability, 215–16
- block diagram, 216, 217, 218
- video capability, 215

Digital video interfaces, 78, 80
- consumer component interfaces, 110–19
- consumer transport interfaces, 119–23
- IC component interfaces, 107–10
- pro-video component interfaces, 81, 84–94
- pro-video composite interfaces, 94–105
- pro-video transport interfaces, 105–7

Digital video processing, 125
- chroma keying, 131–2, 133, 134
- DCT-based compression, 147–50
- display enhancement, 126–30

Digital video processing (*continued*)
interlaced-to-noninterlaced conversion, 144–7
noninterlaced-to-interlaced conversion, 143–4
processing definitions, 125
scan rate conversion, 140, 142–3
video mixing and graphics overlay, 130–1
video scaling, 132, 135–40
Digital Visual Interface (DVI), 64, 110–14
Directed Channel Change Selection Code Table (DCCSCT), 208
Directed Channel Change Table (DCCT), 207
Display Data Channel (DDC), 111, 117
arriving request descriptor, 208
departing request descriptor, 209
Display enhancement, 126
blue stretch, 129
brightness, contrast, saturation and hue controls, 126, 127
color correction, 130
color transient improvement, 126, 128
dynamic contrast, 130
green enhancement, 129
luma transient improvement, 126
sharpness, 126, 128–9
Display scaling, 137–40, 141, 142
Dolby® Digital, 208, 213
Dolby® Digital Plus audio, 208
Dr information, 161
DTCP (Digital Transmission Copy Protection), 120
Dual-Prime Motion Compensation Option, 186
DVD players, 8, 9
DVI, *see* Digital Visual Interface
Dynamic contrast, 130

E

EDTV sample rate selection, 35
End of active video (EAV), 80
Enhanced-definition video, 7
Enhanced signaling descriptor, 209
Enhanced television programming, 162
transports, 163
triggers, 163
Entropy coding, 199
Ethernet, 120
Euroconnector, 65
European Broadcasting Union (EBU), 32
Event information table (EIT), 207
Extended channel name descriptor, 209
Extended Display Identification Data (EDID), 111
Extended Profile (XP), 197
Extended text table (ETT), 207

F

F (field) signals, 38, 40
Field merging, 146
Field rate conversion, *see* Scan rate conversion
Film mode deinterlacing, 144
Fine Granularity Scalable objects, 190
Flexible Macroblock Ordering (FMO), 198
Fractional ratio interpolation, 145
Frame/field dropping and duplicating, 142–3
Frame rate conversion, *see* Scan rate conversion

G

Gamma correction, 26–9
Genre descriptor, 209
Gigabit Video Interface (GVIF), 117–19
Graphics profiles, 192
Green enhancement, 129
Group of pictures (GOP), 172, 186
Group of video object plane (GOV), 193–4

H

H (horizontal blanking) signals, 38, 40
H.264, *see* MPEG-4.10 video

HDMI, *see* High-Definition Multimedia Interface
HDTV (High-Definition Television), 31
HDTV RGB interface, 67–8, 70
HDTV sample rate selection, 35
HDTV video systems, 28
HDTV YPbPr interface, 71–2, 74, 75, 76, 77
High Data-Rate Serial Data Transport Interface (HD-SDTI), 106–7
High-Definition Multimedia Interface (HDMI), 63, 114–17
High-definition video, 7
High Profiles (HP), 197–8
HLS (hue, lightness, saturation) color space, 19
Horizontal sync (HSYNC), 5, 64
HSI (hue, saturation, intensity) color space, 19–22
HSV (hue, saturation, value) color space, 19–22
HTML 4.0, 162
Hue, definition of, 154
Hue control, 126

I

I (intra) pictures, 185
I (intra) frames, 171
I bandwidth, 152, 153, 154
IC component interfaces:
  BT.601 video interface, 108
  BT.656 Interface, 109
  RGB values, 108
  Video Interface Port (VIP), 109–10
  Video Module Interface (VMI), 108
  YCbCr values, 107–8
  Zoomed Video Port (ZV Port), 109
IEEE 1394, 120–1, 122
Inter-field processing, 146–7
Interlaced analog component video, 35
  480i and 480p system, 35–6
  576i and 576p system, 44
  1080i and 1080p system, 54
Interlaced analog composite video:
  480i and 480p system, 35
  576i and 576p system, 41–3
Interlaced digital component video:
  480i and 480p system, 36–40
  576i and 576p system, 45–7
  1080i and 1080p system, 55–6, 57, 58
Interlaced displays, 6
Interlaced-to-noninterlaced conversion, 144
  film mode deinterlacing, 144
  video mode deinterlacing, 144–7
Interlaced video, 170
Intra-field processing, 145
IPMP (Intellectual Property Management and Protection), 175
IPTV set-top boxes, 2
IRE, 64
ISDB digital television, 216, 219
  audio capability, 222
  block diagram, 221, 222
  ISDB-C (cable), 216
  ISDB-S (satellite), 216
  ISDB-T (terrestrial), 219
  video capability, 219–20
ITU BT.709, 35
ITU BT.1358, 35
ITU-R BT.709, 55, 56

J

Javascript 1.1, 162
JPEG vs. MPEG, 166–8

L

Linear interpolation, 136, 137
Lossless compression, 165
Low-voltage differential signaling (LVDS) link, 117, 118
LSB toggling, 149
Luma transient improvement, 126
Luminance information:
  NTSC, 152
  PAL, 156–7
  SECAM, 160
Luminance modulation, 132

M

M-JPEG (motion JPEG), 166, 167
Macroblocks, 172–3, 198
  in MPEG-2, 187
Main objects, 190
Main Profile (MP), 197
Master guide table (MGT), 207
Mini-D ribbon (MDR) connector, 116
Mobile video receivers, 2
Modulator, definition of, 151
Monochrome luminance, 152
Motion adaptive deinterlacing, 146–7
Motion-compensated deinterlacing, 147
Motion compensation:
  for H.264, 198
  for MPEG-1, 172–3
  for MPEG-2, 186
Motion vector steered de interlacing, *see* Motion-compensated deinterlacing
MPEG, 1, 165
  vs. JPEG, 166–8
MPEG-1, 165, 166
  audio coding, 168–9
  mismatch control, 149
  quality issues, 168
  video bitstream, 173–4
  video coding, 169–73
  video decoding, 174
MPEG-2, 2, 8, 165, 174–5
  audio coding, 175
  levels, 176, 177
  profiles, 176–7, 178, 180
  scalability, 177, 179–80, 185
  transport stream, 202, 214
  video coding, 185–7
  video decoders, 180, 187
  video overview, 176–7, 177–9, 180, 181–4
MPEG-4, 165, 187–8
  audio codecs, 188–9
  graphics profiles, 192
  visual codecs, 189–91
  visual layers, 192–5
MPEG-4.2 natural visual object types, 190–1
MPEG-4.10 video, 8, 149, 165
  levels, 196
  profiles, 195, 197–8
  video coding layer, 198–9
Multiple reference frames, 198

N

National Television System Committee (NTSC), 22, 27–8, 28, 31, 151
  color information, 152–3
  color modulation, 153–4
  composite video generation, 154–5
  luminance information, 152
  standards, 155–6
  video timing, 95, 96, 97–9, 100
"Nearest neighbor" scaling, *see* Pixel dropping and duplication
Non-CRT displays, 29
Noninterlaced displays, *see* Progressive displays
Noninterlaced NTSC, 156
Noninterlaced PAL, 159
Noninterlaced-to-interlaced conversion, 143
  scan line decimation, 143–4
  vertical filtering, 144
Non-RGB color space, 25–6
NTSC 4., 43, 155

O

Oddification, 149
Open LVDS Display Interface (OpenLDI), 117, 118
OpenCable™, 210
  audio capability, 213
  block diagram, 213, 214
  video capability, 213
Out-of-Band (OOB) Forward Data Channels (FDC), 210, 213
Out-of-Band (OOB) Reverse Data Channels (RDC), 210, 213

P

P (predicted) frames, 171
P (predicted) pictures, 185, 187
PAL system, *see* Phase Alternation Line system
PALplus, 160
Parallel interfaces, 101, 104
  25-pin parallel interface, 86, 87, 88–91
  93-pin parallel interface, 91–3
Pedestal, 32
Peritel, 65
Phase Alternation Line (PAL) system, 28, 31, 156
  color information, 157
  color modulation, 157–8
  composite video generation, 158–9
  luminance information, 156–7
  PALplus, 160
  standards, 159
  video timing, 95–6, 99, 101
Pixel dropping and duplication, 135–6
Portable media players, 2
Program and system information protocol (PSIP), 203–6
  descriptors, 208–9
  optional tables, 207–8
  required tables, 207
Program stream, 185
Progressive analog component video:
  480i and 480p systems, 36
  576i and 576p systems, 44
  720p system, 48, 51
  1080i and 1080p systems, 54–5
Progressive digital component video:
  480i and 480p systems, 40–1
  576i and 576p systems, 47–8, 49–51
  720p system, 51–5
  1080i and 1080p systems, 56, 57, 58–60
Progressive displays, 5
Pro-video analog interfaces, 77–8, 80, 81, 82, 83
Pro-video component interfaces, 85
  ancillary data, 86, 88
  parallel interfaces, 86–93
  serial interfaces, 93–4, 95
  video timing, 81–6, 87
Pro-video composite interfaces, 94
  ancillary data, 100–1, 102–3
  NTSC video timing, 95, 96, 97, 98, 99, 100
  PAL video timing, 95–6, 99, 101
  parallel interfaces, 101, 104
  serial interface, 104–5
Pro-video transport interfaces:
  High Data-Rate Serial Data Transport Interface (HD-SDTI), 106–7
  Serial Data Transport Interface (SDTI), 105–6

Q

Q bandwidth, 153, 154
QAM (quadrature amplitude modulation) encoding, 201
Quantization, 149, 199

R

Rating region table (RRT), 207
Real Time Interface (RTI) extension, 175
Red, green, and blue (RGB) color space, 15–16
Redistribution control descriptor, 209
RGB chromaticities, 24, 25
RGB values, 108
Run length coding, 150

S

S-video, 3
  interface, 64–5
Satellite VCT (SVCT), 207
Saturation control, 126
Scalability, 177
  data partitioning, 180
  SNR scalability, 179
  spatial scalability, 179
  temporal scalability, 180
  transport and program streams, 185

Scan line decimation, 143–4
Scan line duplication, 145
Scan line interpolation, 145
Scan lines, 5
Scan rate conversion, 140
frame/field dropping and duplicating, 142–3
temporal interpolation, 143
SCART interface, 65, 66, 67
Scrambler circuit, 94
SDTV (Standard Definition Television):
definition, 31
sample rate selection, 33–5
SDTV RGB interface, 65, 67, 68, 69
SDTV YPbPr interface, 69–71, 72, 73, 74
SECAM, 28, 31, 160
color information, 161
color modulation, 161
composite video generation, 161
luminance information, 160
standards, 161–2
Serial Data Transport Interface (SDTI), 105–6
Serial interface, 104–5
Service Location Descriptor, 207, 209
Shadow chroma keying, 132
Sharpness, 126, 128–9
Simple objects, 190
SMPTE 125M, 88
SMPTE 267M, 36–7
SMPTE 274M, 55, 56
SMPTE 293M, 40–1, 42, 43
SMPTE 296M, 51
SMPTE 363M, 162
SMPTE 421M (VC-1), 8, 195
SNR Scalability, 179
Society of Motion Picture and Television Engineers (SMPTE), 32
Spatial scalability, 179
Square pixels, 32
Standard-definition video, 7
Standards organizations, 10
Start of active video (SAV), 80, 84, 86
Subcarrier, definition of, 151
Sync signal, 64
System Renewability Message (SRM) reference descriptor, 209
System time table (STT), 207

T

Temporal interpolation, 143
Temporal rate conversion, *see* Scan rate conversion
Temporal scalability, 180
Terrestrial virtual channel table (TVCT), 207
Time-Shifted service descriptor, 209
Transition-minimized differential signaling (TMDS) links, 111–12
Transport stream, 185
Transport/Transmission System ID (TSID), 207
Transports, 163
Triggers, 163

U

U bandwidth, 152, 153, 154, 157
Universal VLC (UVLC), 199
USB (Universal Serial Bus) 2.0, 119–20

V

V (vertical blanking) signals, 38, 40
V bandwidth, 152, 153, 154, 157
Variable interpolation, 145
Variable-length coding (VLC), 150, 199
VBI, *see* Vertical blanking interval
Vertical blanking interval, 63
Vertical filtering, 144
Vertical sync (VSYNC), 4, 64
VGA interface, 78, 83, 84
Video, definition of, 1
Video bitstream:
MPEG-1, 173–4
Video capability, 203
ATSC digital television, 203
DVB digital television, 215
ISDB digital television, 219–20
OpenCable™, 213

Video coding layer, of H.264
entropy coding, 199
macroblocks, 198
motion compensation, 198
multiple reference frames, 198
transform, scaling, and quantization, 198–9
YCbCr color space, 198
Video coding layer, of MPEG-1, 169
coded frames, 171–2
encoding, 170–1
interlaced video, 170
motion compensation, 172–3
Video coding layer, of MPEG-2:
coded pictures, 185–6
macroblocks, 187
motion compensation, 186
YCbCr Color Space, 185
Video data, 3–4
Video data formats, 108
Video decoding:
MPEG-1, 174
MPEG-2, 180, 187
Video Interface Port (VIP), 109–10
Video interfaces:
analog video interface, 63–78
digital video interfaces, 78–123
Video mixing and graphics overlay, 130–1
Video mode deinterlacing, 144
inter-field processing, 146–7
intra-field processing, 145
Video Module Interface (VMI), 108, 109
Video object (VO), 193
Video object layer (VOL), 193
Video object plane, 194
Video resolution, 6–7, 135
Video scaling, 132
anti-aliased resampling, 136–7
display scaling examples, 137–40, 141, 142
linear interpolation, 136, 137
pixel dropping and duplication, 135–6
Video signals:
1080i and 1080p system, 54–60
480i and 480p system, 35–41
576i and 576p system, 41–8
720p system, 48–53
definition, 31
digital component video, 32–5
Video timing, 4–6, 81–6, 87
NTSC, 95, 96, 97, 98, 99, 100
PAL, 95–6, 99, 101
Video today, 2–3
Visual layers, of MPEG-4:
BIFS, 194–5
group of video object plane, 193–4
video object, 193
video object layer, 193
video object plane, 194
visual object sequence, 193
Visual object planes, 189–90
Visual object sequence (VOS), 193
VMI, *see* Video Module Interface
VRML vs. BIFS, 194

Y

YCbCr color space, 18–19, 25, 198
MPEG-2 video coding, 185
MPEG-4 video coding, 189
YCbCr values, 107–8
YIQ color space, 17–18
YPbPr video signal, 3
YUV color space, 16–17

Z

Zig-zag scanning, 149
Zoomed Video Port (ZV Port), 109